THE NATURE OF TEMPORAL (t > 0) SCIENCE

THE NATURE OF TEMPORAL (t > 0) SCIENCE

A Physically Realizable Principle

Francis T. S. Yu

CRC Press
Taylor & Francis Group
Boca Raton London New York

CRC Press is an imprint of the
Taylor & Francis Group, an **informa** business

First edition published 2022
by CRC Press
6000 Broken Sound Parkway NW, Suite 300, Boca Raton, FL 33487-2742

and by CRC Press
4 Park Square, Milton Park, Abingdon, Oxon, OX14 4RN

CRC Press is an imprint of Taylor & Francis Group, LLC

ISBN: 978-1-032-22194-6 (hbk)
ISBN: 978-1-032-22151-9 (pbk)
ISBN: 978-1-003-27150-5 (ebk)

DOI: 10.1201/9781003271505

Typeset in Times
by MPS Limited, Dehradun

Dedication

---

*It is to discover but not to create the law of nature.
Science is supposed to be approximated, it cannot be virtual and deterministic as mathematics is.*

Laws, principles, and theories were made to be broken, revised, or even eradicated.

Everything within our universe has a necessary price; a section of time Δt and an amount of energy ΔE and it is not free.

Within the quantum field regime, thing changes at the speed of light, but we cannot change the pace of time.

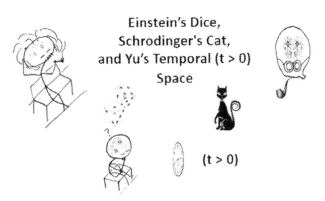

Einstein's Dice,
Schrodinger's Cat,
and Yu's Temporal (t > 0)
Space

(t > 0)

The Art of A Physically Realizable Principle

Contents

Preface

Before I get started, a couple of remarks that I would like make: Firstly, the scientists' dilemma is whether should we bury ourselves within a timeless (t = 0) science and thrive or change to temporal science in bright. Secondly, as a learner, we gradually lose the independent logical thinking and have opted to accept the approval of the others.

One of the objectives for writing this book is to show that we had been inadvertently burying ourselves within a timeless (t = 0) paradigm for centuries, although deterministic laws, principles, and theories were still the cornerstones of our science. But some of them are as fictitious and virtual as mathematics is, which did not actually exist within our temporal (t > 0) universe. For example, Schrödinger's fundamental principle of superposition and Einstein's special theory of relativity and his general theory of relativity did not exist within our temporal (t > 0) space. In this book I will show Schrödinger's fundamental principle of superposition is not a physically realizable principle within our temporal (t > 0) universe. Secondly, I will show that Einstein's special theory of relativity is also not a physically realizable theory, and his general theory of relativity should be a non-deterministic principle since within our temporal (t > 0) universe; future prediction cannot be deterministic because science is supposed to be approximated. Although quantum and relativistic mechanics revolutionized our classical mechanics for over a century, at the same time they have created one of the greatest science conspiracies.

Since science needs mathematics, but mathematics is not equal to science, we have seen that mathematics overshadowed the significant of physical realizability of science. From this we have seen a score of theoretical analyses and principles that are fictitious and virtual and are not actually physically realizable within our temporal (t > 0) space. Recently I have discovered the origin of all non-physically realizable principles and theory, which was from the background of a piece of commonly used paper for scientifical model analyses that have been used for centuries, but not knowing that an empty space model is not a physically realizable paradigm within our universe since our universe has time. From this, we see that it is not how rigorous mathematical analysis is, it is the physically realizable paradigm that determines the physically realizable science.

I had recently found that our universe is a temporal (t > 0) universe, of which time is a dependent variable within our universe, instead of a four-dimensional space-time of Einstein as commonly used. The major difference is that our temporal (t > 0) time-space is space changes with time, but not space changes space-time. In other words, every subspace within our temporal (t > 0) universe is temporal, where time is a dependent variable coexists with space. From this it is very appropriate for me to begin Chapter 1 on the nature of temporal (t > 0) space, where every physically realizable science has to comply.

Nevertheless, one of the most important aspects within a temporal (t > 0) universe is that every subspace (i.e., substance or matter) has a price to pay; namely, a section of time Δt and an amount of energy ΔE (i.e., Δt, ΔE) that includes our universe, and it is not free. As in contrast with the commonly used zero-summed energy spacetime continuum which has no prize to pay.

One of the most important divisions between physical science and virtual mathematics is that one is required to physically exist within our universe and the other is not required. This must be the major reason to let the scientific community know that a score of non-physically realizable timeless (t = 0) theories and principles have created a worldwide scientific conspiracy. If we do not stop it sooner, it will cause more damage to our science than benefits. Therefore, it is about time for us to look into temporal (t > 0) science; otherwise, we will continually be trapped within an empty land of fantasy science, which does not have any price to pay, because it has no time and no energy to pay.

This book is intended for cosmologists, particle physicists, astrophysicists, quantum physicists, computer scientists, communication engineers, professors, and students as a reference and a research-oriented book.

Author

Francis T. S. Yu received his B.S.E.E. degree from the Mapua Institute of Technology, Manila, Philippines, and his M.S. and Ph.D. degrees in electrical engineering from the University of Michigan. During the period from 1958 to 1965, he was a teaching fellow, an instructor, and a lecturer in the Electrical Engineering Department at the University of Michigan, and a research associate with the Communication Sciences Laboratory at the same university. From 1966 to 1980, he was on the faculty of the Electrical and Computer Engineering Department at Wayne State University. He was a visiting professor in the Electrical and Computer Engineering Department at the University of Michigan from 1978–1979. In 1980, he became a professor in the Electrical Engineering Department at The Pennsylvania State University. He has been a consultant to several industrial and government laboratories. He is an active researcher in the fields of optical signal processing, holography, optics and information theory, and optical computing. He has published over 300 refereed papers in these areas. He is a recipient of the 1983 Faculty Scholar Medal for Outstanding Achievement in Physical Sciences and Engineering, a recipient of the 1984 Outstanding Researcher in the College of Engineering, was named Evan Pugh Professor of Electrical Engineering in 1985 at Penn State, a recipient of the 1993 Premier Research Award from the Penn State Engineering Society, was named honorary professor at Nankai University in 1995, the co-recipient of the 1998 IEEE Donald G. Fink Prize Paper Award, named Honorary Professor in National Chiao Tung University Taiwan in 2004, the recipient of the 2004 SPIE Dennis Gabor Award, and the 2017 OSA Emmet N. Leith Medal. Yu is a life-fellow of the IEEE and fellow of OSA, SPIE, and PSC. He retired from Penn State University in 2004. He is the author and co-author of 13 books entitled: (1) *Introduction to Diffraction, Information Processing and Holography* (translated in Russian), (2) *Optics and Information Theory*, (3) *Optical Information Processing* (translated in Chinese), (4) *White-Light Optical Signal Processing*, (5) *Principles of Optical Engineering* (with I. C. Khoo) (translated in Chinese), (6) *Optical Signal Processing, Computing, and Neural Networks* (with S. Jutamulia) (translated in Chinese and Japanese), (7) *Introduction to Optical Engineering* (with X. Yang) (translated Korean), (8) *Entropy and Information Optics* (translated in Chinese), (9) *Introduction to Information Optics* (with S. Jutamulia and S. Yin) (translated in Chinese), (10) *Coherent Photonics* (in Russian) (with A. Larkin in Russian), (11) *Neural Networks and Education: The Art of Learning* (translated in Chinese, Spanish and Russian), (12) *Neural Stickman: The Art of ...*(translated in Chinese, Spanish and Russian), and (13) *Origin of Temporal (t > 0) Universe: Connecting to Relativity, Entropy, Communication, and Quantum Mechanics*. And he also has contributed several invited chapters in various monographs and books. He has co-edited four books entitled: (1)*Optical Storage and Retrieval* (with S. Jutamulia), (2) *Optical Pattern Recognition* (with S. Jutamulia), (3) *Photorefractive Optics* (with S. Yin), and (4) *Fiber Sensors* (with S. Yin). He has also co-edited two volumes of the SPIE Milestone Series: *Optical Pattern Recognition* (with S. Yin) and*Coherent Optical*

Processing (with S. Yin). And chairs/editors (with R. Guo and S. Yin) over 25 volumes of SPIE Proceedings on Photorefractive Fiber and Crystal Devices: Materials, Optical Properties, and Applications. Yu's most notable work must be his current book on *Origin of Temporal (t > 0) Universe: Connecting to Relativity, Entropy, Communication, and Quantum Mechanics* and his article on "What is Wrong with Current Theoretical Physics". Currently, he is one of the advisory editors for *Asian J Phys*.

1 Nature of Temporal (t > 0) Space

Since science needs mathematics, but we should not let mathematics take over our science since mathematics is not equal to science. It is therefore our responsibility to produce science that is physically realizable rather than fictitious and virtual, as is mathematics. Since science depends on mathematics, then what would you anticipate from science if it is "strictly" from mathematics standpoint? Since science is a law of "approximation", in which we see that without approximated science then there will be no science since science is not supposed to be exact. On the other hand, what about mathematics? The answer is that mathematics is an axiom of absolute "certainty". For this we see that, dependency on exact mathematics to evaluate approximated science is anticipated as "inexact". Since science has to be physically real and mathematics is virtually rigorous, we see that it is "not" how sophisticated or rigorous mathematics is, it is how mathematics relattes to science. For this we see that it should be that science directs the mathematics, but not letting mathematics take over the leadership of science.

In order to understand the physical realizable science, first of all we have to understand what kind of a physical subspace we are living in. For example, when we hypothesize a deep-sea submarine, we need to understand the possible scenarios of the undersea environment, otherwise the submarine that we are hypothesizing would not physically realizable. For which it is my privilege to introduce how our temporal (t > 0) universe was created, which is a physically realizable subspace that all physical sciences are situated within. Otherwise, any hypothetical science would be as virtual as mathematics is. In other words, any scientific law, principle, theory, and hypothesis has to be proven existed within our temporal (t > 0) universe, otherwise they cannot actually exist within our temporal (t > 0) universe. From this, I have the fundamental principle of temporal (t > 0) universe to share; one cannot not get something from nothing within our universe there is always a price to pay, which is a section of time Δt and an amount of energy ΔE and it is not free. From which I will show that, our universe is an energy conservation subspace, and it is not a zero-summed energy universe that commonly had assumed.

1.1 TEMPORAL (t > 0) SUBSPACE

In view our universe, time must be one of the most enigmatic variables in the law of science. So what is time? Time is a variable and not a substance. It has no mass, no weight, no coordinate, no origin, and it cannot be detected or even be seen. Yet time is an everlasting variable within our known universe; it has no beginning and no end. From this we see that without time there would be no physical substance, no

DOI: 10.1201/9781003271505-1

physical space, and no life. The fact is that every physical substance coexists with time, including our universe. Therefore, when one is dealing with science, time is one of the most enigmatic variables that cannot be simply ignored. Strictly speaking, all the laws, principles, and theories of science as well all the physical substances cannot exist without the existence of time.

On the other hand, energy is a physical quantity that governs every existence of substance, which includes the entire universe. In other words, without the existence of energy, there would be no substance and no universe! Nonetheless, based on our current laws of science, all the substances were created by energy and every substance can also be converted back to energy. Thus, energy and substances are exchangeable, but it requires some physical conditions (e.g., nuclei and chemical interactions and others) to make the conversion start. Since energy can be derived from mass, mass is equivalent to energy, yet mass is not equal to energy. Hence, every mass can be treated as an energy "reservoir". The fact is that our universe is compactly filled with mass and energy. Without the existence of time, the trade (or conversion) between mass and energy could not have happened.

Let us now start with Einstein's energy equation, which was derived from his special theory of relativity [1], as given by:

$$E \approx Mc^2 \qquad\qquad (1.1)$$

where "$\approx$" is an approximation sign since science is approximated, m is the rest mass, and c is the velocity of light. In view of Eq. (1.1), we see that it is a point-singularity approximated timeless (t = 0) or time-independent equation. In other words, the equation needs to convert into a temporal (i.e., t > 0) or time-dependent equation for the conversion to take place from mass into energy. For which we see that, without the inclusion of time variable, the conversion would not have had taken place. Nonetheless, Einstein's energy equation represents the total amount of energy that can be converted from a rest mass, M. From this we see that every mass can be viewed as an energy reservoir. Thus, by incorporating a time variable, Einstein's energy equation can be represented by a partial differential form, as given by [2,3]:

$$\frac{\partial E(t)}{\partial t} \approx c^2 \frac{\partial M(t)}{\partial t}, \quad t > 0 \qquad\qquad (1.2)$$

where $\partial E(t)/\partial t$ is the rate of increasing energy conversion, $(-\partial M(t)/\partial t)$ is the corresponding rate of mass reduction, c is the speed of light, and t > 0 represents a forward time-variable or equation that exists only in the positive time domain. In this we see that Eq. (1.2) is a time-dependent equation that exists in time t > 0, which represents a forward time variable thay only occurs after excitation at t = 0. Incidentally, this is the well-known causality constraint (i.e., t > 0) [4] imposed by our universe.

Since light velocity is dependent on permittivity and permeability (i.e., μ ε) within a medium that light travels, from this we see that light speed "cannot" be constant as normally assumed it is constant. But it is more important to tell us that without a medium (i.e., substance), the light wave cannot travel within an empty space. From this we see that space beyond our universe is "not empty" and it has a similar substance (i.e., $\mu\ \varepsilon$); otherwise, our universe cannot be a "bounded" sub-space [2,3] for which the boundary is limited by the speed of light. Since it was speculated that the boundary of our universe expands fasters [5], and as I see it, it must be the radial velocities of substances (i.e., subspaces) are linearly increasing as the boundary expands at the speed of light. In this we see that substances near the boundary moves faster, as we view the substances near the boundary of our universe. It is trivial that we will never be able to see that boundary of our universe since the boundary is expanding at the speed of light.

Nevertheless, one of the important aspects of Eq. (1.2) must be that energy and mass can be traded, for which the rate of energy conversion from a mass can be written in terms of electromagnetic (EM) radiation or radian energy as given by [6]:

$$\frac{\partial E}{\partial t} = -c^2\frac{\partial M}{\partial t} = [\nabla\cdot S(\upsilon)] = -\frac{\partial}{\partial t}\left[\frac{1}{2}\varepsilon_0 E^2(\upsilon) + \frac{1}{2}\mu_0 H^2(\upsilon)\right], \quad t > 0 \quad (1.3)$$

where ε_0 and μ_0 are the permittivity and the permeability of the deep space, re-spectively; υ is the radian frequency variable; $E^2(\upsilon)$ and $H^2(\upsilon)$ are the respective electric and magnetic field intensities; the negative sign represents the outflow of energy per unit time from a unit volume; $(\nabla\cdot)$ is the divergent operator; and S is known as the Poynting vector or energy vector of an electromagnetic radiator as given by:

$$S(\upsilon) = E(\upsilon) \times H(\upsilon) \qquad (1.4)$$

where $E(\upsilon)$ and $H(\upsilon)$ are electric and magnetic field intensity, respectively; and x is the cross product. From this we see that Eq. (1.3) is a point-singularity approxi-mated "time-dependent" equation; $t > 0$ denotes that equation is existed within the positive time domain or temporal (t > 0). From this we see that radian energy (i.e., radiation) diverges from a reducing mass, as mass annihilates with time. In other words Eq. (1.13), is not just a mathematical formula but it shows the transformation from a point-singularity approximation mass M to a three-dimensional space-time representation. For this we see that the boundary is continually expanding with time, where time is a forward-dependent variable with respect to the creating space. Since empty and non-empty space cannot coexist, our universe had to be created within a non-empty space, as depicted in Figure 1.1(a), instead of an empty space as normally assumed, in Figure 1.1(b) [7].

In view of the big bang creation theory [7], I see that the big bang creation was started within a preexistent temporal (t > 0) space, as shown in Figure 1.1(a); otherwise, the creation cannot get started as had been hypothesized within a timeless (t = 0) space, as depicted in Figure 1.1(b). But it is not a physically rea-lizable paradigm since non-emptiness cannot exist within an empty space, which is known as the temporal (t > 0) principle. Secondly, speed of radiation will be

(a)   (b)

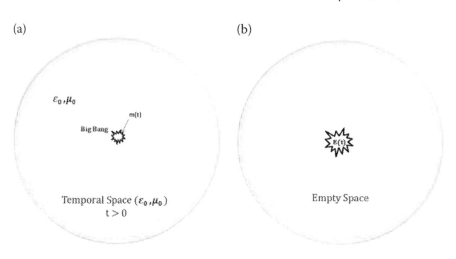

FIGURE 1.1   Shows a set of postulated big bang creation paradigms. Figure 1.1(a) shows a big bang creation started within a physically realizable temporal (t > 0) space, where time had have existed within the space well before the creation. Figure 1.1(b) shows a big bang creation started within a timeless (t = 0) empty space, which it is not a physically realizable paradigm but had have been commonly used.

unlimited even though we ignored the physically realizable issue because empty space has no substance in it. As a matter of fact, empty space has no time and no space for which we see that empty space is virtual mathematical space that scientists had been using for centuries since the dawn of science [8].

Yet, within our universe, every matter is a substance, which includes all the fundamental particles, electric, magnetic, gravitation fields, and energy. The reason is that they were all created by means of energy or mass. The fact is that our physical space (e.g., our universe) is fully "packed" with substances (i.e., mass and energy) and left "no" absolute empty subspace within it, since emptiness and non-empty are mutually exclusive. For this I stress that empty space is not an in-accessible subspace within our universe, as in contrast with some scientists who believe that empty spaces are accessible within our universe. In fact, empty space is a virtual mathematical abstract space and it does not exist within our universe since every subspace within our universe cannot be empty. From this we see that sub-stance within our universe cannot just be limited by particles (e.g., permeability and permittivity and others).

In view of the big bang creation formular of Eq. (1.3), we see that it is a sto-chastic space-time dynamic equation that constantly changes with time. For this we see that every created subspace within the created universe is a non-empty subspace that changes with time, which includes the created universe itself. From this we see that it is not possible for any subspace or our universe to change the time or curves the space time. For example, we change with time, but we cannot change time. Since time is a dependent variable, it moves at a constant pace and coexists with space or substance, but time has no beginning and no end. In other words, there will always be space although every substance or subspace (e.g., our universe) has a beginning and an end, but time has no beginning and no end. In this we see that every amount of energy (i.e., substance or subspace) changes with time (i.e., $\Delta E(t)$). Since every substance or subspace has a life (i.e., a section of time Δt), then we see that every subspace or substance is limited by a section of time Δt and an amount of energy $\Delta E(t)$, while $\Delta E(t)$ changes with time. In this we see that we can change the section of time Δt as small as we wish (i.e., $\Delta t \to 0$) but we cannot squeeze $\Delta t = 0$, even though we assumed we had all the energy $\Delta E(t)$ we needed.

Since no physical substance within our universe can exist forever or without time, this includes our universe. Thus, without time there would be no substance and no universe because every physical substance described itself as a physical space and it is changing naturally with time. In this we see that every physical substance is itself a temporal (t > 0) subspace, which includes our universe and is called the temporal (t > 0) universe. In other words, every temporal (t > 0) subspace is filled with a substance that changes naturally with time. From this, every sub-space cannot be empty (even one or two dimensional) subspace since every sub-space has to have substance (or energy) in it to be qualified as a temporal (t > 0) subspace. Therefore, no matter how small a substance (or particle) is, it has to be temporal (t > 0) (i.e., existed with time), otherwise it cannot exist within our universe. Since our universe has time or is temporal (t > 0), this means it is physically realizable only in the positive time domain t > 0 (i.e., future time domain). And it does not exist in the negative time domain t < 0 (i.e., past time domain) since time is no longer there. From this we see that the present moment (i.e., t = 0) is the only absolute moment of physical reality. In other words, the present moment (t = 0) is the only moment of the truth physically real moment. Although the past time do-main (t < 0) was the precise certain moment, it has no physical substances, no real time, and it is virtual like mathematics. Yet the future moment (i.e., t > 0) is the physically realizable moment but uncertain, which means that the future moment cannot be predicted with absolute certainty. In other words, the further away from the present moment (t = 0), the more difficult it is to predict and more uncertain. In this we see that every physically realizable subspace, substance, or science cannot predict with absolute certainty. From this we see that science is not supposed to be exact or deterministic.

Yet without the past time (i.e., t < 0), certainties of precise memories or past-time universes, it has no better way to predict our future universe or science. And this is precisely why all the laws, principles, and theories are deterministic. This is precisely the reason we demand laws, principles, and theories to be exact or deterministic, and suddenly we found they fail to exist within our time-changing universe. Schrödinger's fundamental principle of quantum mechanics [9], Einstein's special and general theory of relativity [1], and many others that I had found failed to exist within our temporal (t > 0) universe. It is because our universe changes with time, but all the laws, principles, and theories cannot change with time since they are all deterministic principles. Yet all the laws, principles, and theories are supposed to be non-deterministic or uncertain.

Nevertheless, let me epitomize our universe in a diagram, as shown in Figure 1.2. In this we see that it is the present moment t = 0 that divides the past-time (t < 0) virtual domain of precise certainty and future-time (t > 0) physically realizable domain of

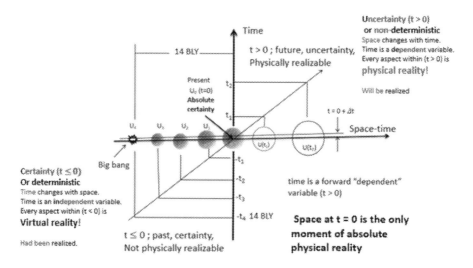

FIGURE 1.2 Shows a composited temporal (t > 0) time-space diagram to epitomize the nature of our temporal universe. BLY is billion of light years. Instant-present moment t = 0 shows the only moment of absolute physically certainty of our universe. Past-time domain (t < 0) shows past certainty universes but without real time and no physical substance. And future-time domain (t > 0) represents a physically realizable domain that changes with time.

uncertainty. In this we see that our universe (i. e., t > 0) changes with time. Since Schrödinger's fundamental principles of superposition demands stopping the clock (i.e., t = 0), and Einstein's special theory of relativity demands the clock to tick ahead or delay the pace of time (i.e., $t = 0 + \Delta t$ or $t = 0 - \Delta t$), our temporal (t > 0) universe cannot change the time although it changes with time. For this it is not possible for Schrödinger to stop the clock and it is also impossible for Einstein to move ahead or behind the speed of time.

Nevertheless, it is not possible to squeeze a section of Δt to zero as Schrödinger fundamental principle of superposition tells us it could stop the clock as given by:

$$[t = 0 + \Delta t; \quad \Delta t = 0], \quad \Rightarrow t = 0 \tag{1.5}$$

which is essentially demanding time to stop before it ticks? But as time stands still at t = 0, it represents a timeless (t = 0) space that has no time and no space. Then all particles will be superimposing together simultaneously and instantaneously as Schrödinger's superposition principle shows it could, since the principle was derived within an empty timeless space. But empty space cannot exist within our time-space, so time did not stop, by which we have the following relationship:

$$[t = 0 + \Delta t; \quad \Delta t \to 0], \quad \Rightarrow t = 0 + \Delta t \tag{1.6}$$

From this we see that Schrödinger's fundamental principle of superposition collapses or "fails" to exist within our temporal (t > 0) universe since particles are very unlikely to be simultaneously and instantaneously superimposing together at any time.

Similarly, Einstein's special theory of relativity shown that moving particle can move ahead or behind the pace of time as given by:

$$[t = 0 + \Delta t \text{ or } t = 0 - \Delta t], \quad \Rightarrow t = 0 + \Delta t \text{ or } t = 0 - \Delta t \tag{1.7}$$

in which Einstein's special theory demands that time moves forwardly or backwardly beyond or behind the pace of time at $t = 0$. From this we see that Einstein can stop the clock momentarily to let a particle move ahead or behind the pace of time. But the pace of time cannot be slowed down or be sped up, for which we have the relationship as given by:

$$[t = 0 + \Delta t \text{ or } t = 0 - \Delta t], \quad \Rightarrow t = 0 \tag{1.8}$$

From this we found Einstein's special theory fails to exist within our temporal $(t > 0)$ universe. Nevertheless, physical reality occurs if and only if at current time of certainty (i.e., $t = 0$). In other words, physical realities were existed once at precise past time (i.e., $-t_4 = -14\text{BLY}$, $-t_3$, $-t_2$, $-t_1$, $t = 0$) of certainty shown in the figure. From which we see that all those past $(t < 0)$ certainties were virtual memories with no real physical substances. While future $(t > 0)$ [i.e., $(t = 0 + \Delta t)$] will be physically realized but with degree of uncertainty. In other words, future is always unpredictable, the further away from currently certainty the more uncertain to predict. From we see that, the law of nature is not supposed to be exact, yet we want our science to be exact.

1.2 EMPTY SPACE

An absolute empty space has no time, no substance, no coordinate, and cannot be event bounded or unbounded. It is a virtual mathematical timeless space (i.e., $t = 0$) and it does "not" exist within our temporal $(t > 0)$ universe, since empty and non-empty subspaces are mutually excusive. Although empty space has no substance in it, mathematicians "can" implant a coordinate in it, as depicted in Figure 1.3. In this

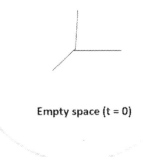

Empty space $(t = 0)$

FIGURE 1.3 Shows an abstract empty space, which is a virtual space that has no time like a piece of plain paper that we commonly use for analyses. Although it has no real time and no real space, scientists can implant fictious time and space in it.

we see that mathematician can do, although it is "not" supposed to have a dimension within an empty pace.

Since scientists are partially mathematicians, no wonder the timeless (t = 0) quantum theory behaves irrationally and that Einstein and his colleagues had objected to the quantum theory severely [10]. But they did not actually pintpoint where the fault of quantum theory comes from since Einstein had committed the same error in where his theory was developed from. (See Chapter 4.)

Since within our physical space it has time and it cannot be empty, this tells us that empty space is an abstract mathematics space, yet we have been using it for centuries. And this is the empty subspace that all of the scientific laws, principles, and theories were developed from this platform, for which all the laws, principles, and theories are timeless (t = 0) or time-independent laws. Strictly speaking, timeless (t = 0) or time-independent laws and theories are fictitious and virtual, although many of them were and "still" are the cornerstones of our science. For example, Einstein's energy equation is one of them, as given by:

$$E = mc^2 \tag{1.9}$$

We see that it is essentially a point-singularity approximated timeless (t = 0) or time-independent equation. However, without the transformation of this equation to a time variable formula, it "cannot" be directly implemented within a temporal (t > 0) subspace since temporal (t > 0) subspace has time. Nevertheless, Einstein's energy equation has given us one of the most important connections between mass and energy, that is energy and mass are equivalent, but not equal.

Since Einstein's energy equation was developed from his special theory of relativity but his special theory is not a physically realizable theory (see Chapter 5), then how can we justify his energy equation is legitimate? In view of Eq. (1.9) we saw that this equation was derived based on kinetic energy dynamics [i.e., $E = (1/2) Mv^2$] where v is the velocity of a mass. For this it is justifiable for us to rewrite his energy equation from the kinetic energy standpoint as given by:

$$E = (1/2)Mc^2 \tag{1.10}$$

From this we see that Eq. (1.10) is consistent with the same physical significance of Einstein; that mass and energy are equivalent.

Although timeless (t = 0) space, Newtonian space [11], and Einstein's space-time [1] had treated time as an "independent" variable, those spaces are virtual spaces that "cannot" exist within our temporal (t > 0) universe since time has to coexist with space. From this we see that the timeless (t = 0) virtual space, Newtonian space, and Einstein's space-time are not "inaccessible" subspaces and cannot exist within our temporal (t > 0) universe; in contrast, as commonly believed, empty space is an inaccessible subspace within our temporal (t > 0) universe. For example, within our temporal (t > 0) universe time-space changes with time, but not time-space curves or changes the space-time, as Einstein's general theory had assumed. Nevertheless, empty space paradigm is a zero-summed energy space as most theoretical physicists had assumed. Which is in fact a

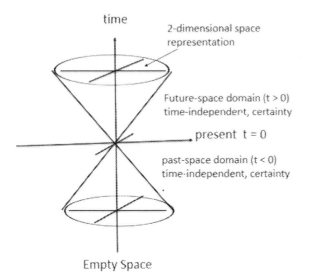

time

2-dimensional space
representation

Future-space domain (t > 0)
time-independent, certainty

present t = 0

past-space domain (t < 0)
time-independent, certainty

Empty Space

FIGURE 1.4   Shows a four-dimensional space-time continuum. It is not a physically realizable paradigm. Yet practically all the theoretical laws, principle, and theories were developed from this subspace paradigm.

four-dimensional space-time continuum that most physicists had used as depicted in
Figure 1.4. And this is precisely the reason theorical physicists had predicted that anti-
matter can exist within our universe (see Chapter 6). But we shall show that, our
universe is in an energy conservation subspace, where time is coexisted with space.
Anti-matter is virtual as mathematics is.

In view of the 4-dimemsional space-time paradigm (see Appendix E), we note that it
is not a physically realizable subspace that should be used in theoretical analyses, since
it violates the second law of thermodynamics where entropy does not increase naturally
with time. It is a time-independent deterministic subspace, as virtual as mathematics is.
But science is supposed to be approximated. And this is precisely the reason, why
quantum theory and relativity principle behaved so weird. Yet we had continuingly
promoting them inadvertently or not let you Judge. But as a scientist, it is our obligation
to bring us back to the physical realizable science rather than bury ourselves within the
fantasy mathematical science for the self-serving of thieve. Otherwise it will be more
damaging to our science than personal gain.

1.3   CREATION OF THE TEMPORAL (t > 0) UNIVERSE

As we accept the big bang theory for our universe's creation [7], then creation
started with Einstein's time-dependent energy formula of Eq. (1.3), as I repeat here
for convenience:

$$\frac{\partial E}{\partial t} = -\left(\frac{1}{2}\right)c^2\frac{\partial M(t)}{\partial t} = [\nabla \cdot S(\upsilon)] = -\frac{\partial}{\partial t}\left[\frac{1}{2}\varepsilon_0 E^2(\upsilon) + \frac{1}{2}\mu_0 H^2(\upsilon)\right], \quad t > 0 \quad (1.11)$$

In this I have used the new energy equation for consistency, where $[\nabla \cdot S(v)]$ represents a divergent energy operation. In this equation, we see that a broad-spectral-band intense radian energy diverges (i.e., explodes) at the speed of light from a point-singularity approximated mass of M(t), where M(t) represents a gigantic mass of energy reservoir. It is apparent that the creation was not ignited by time as commonly believed, but ignited by a huge induced gravitational pressure from mass M(t) itself (see Chapter 5). Furthermore, this equation [I.e., $\nabla \cdot S(v)$] shows that the exploded debris (i.e., matter and energy) starts to spread out in all radiated directions, similar to an expanding air balloon. The boundary (i.e., radius of the sphere) of the universe expands at the speed of light, as the created debris is disbursed. It took about 14 billion chaotic light years [12] to come up with the present observable constellation, but the boundary is still expanding at the speed of light beyond the current observation. This means that the creation process is by no means stopping.

Since our universe was created within a preexistence temporal (t > 0), from this I have no clue to predict when the creation of our temporal (t > 0) will eventually stop, although our universe has a life. This is in contrast with what is commonly believed that our universe will eventually collapse back to a point of singularity. Since our universe was created by means of a big bang explosion within a larger temporal (t > 0) space, it is more logical to anticipate our universe will eventually die off at time approaching to infinity, similar to a circular rain-drop wavelet disappearing on the surface of a water pond. Nevertheless, as I predict that somewhere within the greater temporal (t > 0) space that our universe was embedded, there will be another singularity mass to begin another new universe, which is beyond our current comprehension.

Nevertheless, I stress that one of the important equations showing how our universe was created can be seen in Eq. (1.11), from which we see that a big bang mass annihilation to energy conversion was from a point singularity equation to a space-time creation (i.e., $\nabla \cdot S$). From this we see that Eq. (1.11) is not just a piece of mathematical formula; it is a symbolic representation, a description, a language, a picture, or even a video as may be seen the space is continuingly changes naturally with time. From this we see that our universe is a dynamic stochastic temporal (t > 0) universe, and every consequence within our universe changes naturally with time. In other words, our universe is an energy conservation subspace that expands at speed of light, and it is not a zero-summed energy subspace that allows anti-matter to exist.

1.4   TRADING TIME WITH SUBSPACE

Let us now take one of the simplest connections between physical subspace and time as given by:

$$d = vt \tag{1.12}$$

where d is the distance, v is the velocity, and t is the time variable. Notice that this equation may be one of the most profound connections between time and physical space (or temporal space). Therefore, a three-dimensional time-varying (Euclidean) subspace can be described by:

$$(dx, \ dy, \ dz) = (vx, \ vy, \ vz)t \qquad (1.13)$$

where *(vx, vy, vz)* are the velocity vectors, and t is the time variable. Under the current laws of science, the speed of light is the limit. Then, by replacing the velocity vectors equal to the speed of light, c, a temporal space can be written as:

$$(dx, \ dy, \ dz) = (ct, \ ct, \ ct) \qquad (1.14)$$

Thus, we see that time can be "traded" for space and space "cannot" be traded for time, since time is a forward variable (i.e., t > 0). In other words, once a section of time Δt is expended, we cannot get it back.

Although we have shown a three-dimensional time-varying subspace described by the preceding Eq. (1.14), but the space is still an empty space that is not a physically realizable subspace, since empty and non-empty are mutually exclusive. Then a time-varying space has to be filled with a "time-varying substance". In other words, it is the substance that creates the non-empty time-varying space, or substance. From this we see that time and space are coexisted. Since energy is equivalent to mass and mass is substance, it is an amount of energy ΔE with a section of time Δt that created the subspace. And space (or substance) exists with time, and it is "not" time exists with space. But it is a section of time that exists with space. In this, every subspace within our universe was created by an amount of energy with a section of time [i.e., ΔE(t), Δt]. Thus, we see that the subspace [i.e., E(t)] changes naturally with time but it also coexists with a section of time Δt [i.e., ΔE(t), Δt]. From this we see that every subspace or substance "no" matter how small it is, is a time-energy interdependent subspace [i.e., a temporal (t > 0) subspace], where time is a forward-dependent viable [i.e., (t > 0)]. Every subspace can be represented by an uncertainty relationship as given by:

$$\Delta E \ \Delta t \ \geq \ H \qquad (1.15)$$

where

$$H = (1/2)c^2 \Delta M \ \Delta t \qquad (1.16)$$

where H is least energy $\Delta E = (1/2) (\Delta M) c^2$ to create the substance (or subspace) and ΔM is an amount of equivalent rest mass, by virtue of Einstein's energy equation. From this we see that ΔE and Δt can be mutually traded, which is an amount of energy ΔE that can be traded for a section of time Δt, but it cannot change the speed of time. In other words, the larger a quantity of ΔE gives rise to a narrower section of Δt and vice versa. Nevertheless, the amount of ΔE and the section of Δt are the "necessary cost" for the creation of a specific subspace (or substance). For example, to create a piece of facial tissue, a huge amount of energy

ΔE and a section of time Δt is required to create (i.e., ΔE, Δt), which is the necessary cost. But it is not sufficient yet to make it happen unless an amount of information ΔI or equivalently an amount of entropy ΔS makes it sufficient.

Energy can be presented in various forms such as thermal energy, kinetic energy, potential energy, chemical energy, radiation energy, quantum leap energy, and others. For example, in the form of kinetic energy dynamics, the H quantity in the form of the uncertainty equation can be written as:

$$H = \tfrac{1}{2} \ M \ c^2 \qquad (1.17)$$

In this, M is assumed a quasi-static mass, since within our universe every substance or particle is temporal (t > 0), where c it the velocity of light.

Since quantum mechanics is the legacy of the Hamiltonian classical mechanics [13], we see that the H factor for the quantum uncertainty relationship in Eq. (1.15) reduces to:

$$\Delta E \ \Delta t \geq h \qquad (1.18)$$

which is identical to Heisenberg's uncertainty principle in quantum theory [13], where h is Planck's constant. The major difference of Heisenberg's principle is based on the wave-particle duality of de Broglie's hypothesis [14], which treated a package of wavelet energy equivalents to a particle (i.e., photon) dynamic. However, particle dynamic is "not" equal to a package of wavelet energy [15] (see Appendix B). In this we see that photon is not a particle, but a "virtual" particle has momentum but no mass. This is mainly due to the legacy of quantum mechanics as developed from the classical mechanics standpoint as given by [15,16]:

$$p = c/\lambda \qquad (1.19)$$

where p is the momentum, but it shows no sign of direction, c is the velocity of light, and λ is the wavelength. For this, most of the scientists have treated quantum state energy hυ as a particle (i.e., photon). Similarly, Einstein's energy equation standpoint energy and mass are "equivalent", but energy and mass are "not" equal.

Nevertheless, every subspace within our universe was created by ΔE and Δt, every subspace or substance is a temporal (t > 0) subspace, from which our universe is a huge temporal (t > 0) subspace that exists within an even greater temporal (t > 0) space. Thus, we see that time is a real forward "dependent" variable, but it has "no" beginning and has "no" end. Unlike energy, which is a physical quantity that we can manipulate. In this we see that it is time that we cannot change, but we can change a section of time (i.e., Δt).

Since the creation of our universe was started from a huge explosion of divergent energy (i.e., big bang explosion), as from Eq. (1.11) we see that a spherical temporal (t > 0) universe would be a better presented to describe our universe as given by:

$$r = c \cdot t \qquad (1.20)$$

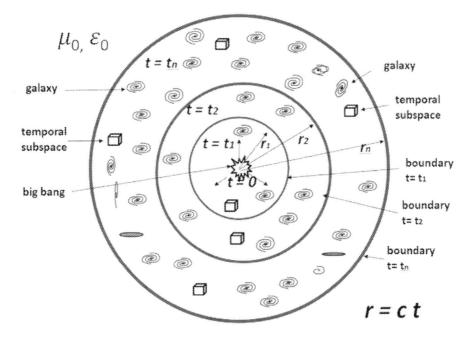

FIGURE 1.5　Shows a composite temporal (t > 0) universe diagram. r = c t, r is the radius of our universe, t is time, c is the velocity of light, ε_0 and μ_0 are the permittivity and permeability of the space.

In this we see that the radius r of our universe increases at the speed of light, where c is the velocity of light and t is time. It is trivial to see that the boundary (i.e., edge) of our universe is determined by radius r, which is limited by the light speed, as illustrated in a composite temporal-space diagram in Figure 1.5. From this we see that the boundary of our universe expands at the speed of light well beyond the current observable range.

From this composite universe diagram, we see that our universe is a dynamic stochastic temporal (t > 0) universe, as depicted in Figure 1.6. From this we see that every aspect within the universe changes "naturally" with time as the boundary expands constantly at the speed of light. For example, the subspace enlarges relatively faster as it is closer toward the boundary, but solid substance m(t) changes little within the subspace. We also see that the outward speed of the subspaces (or particles) increases linearly as the boundary of our universe increases at the velocity of light. For example, outward speed of particle 2 is somewhat faster than particle 1 (i.e., $v_2 > v_1$).

However, our universe is a temporal (t > 0) stochastic universe, which is not an easy task to describe by a simple mathematical equation or geometrical subspaces. But one important aspect of our universe is that every subspace is temporal (t > 0), no matter how small it is. Secondly, subspace "cannot" be empty or time independent. For this, one- or two-dimensional subspaces "cannot" exist within our temporal (t > 0) universe. In view of the current available mathematical abstract

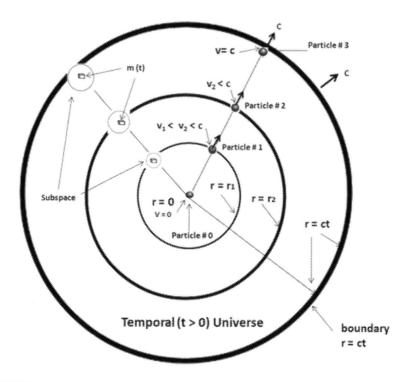

FIGURE 1.6 A schematic diagram of our temporal (t > 0) universe. c is the speed of light, m(t) is the temporal mass, and v is the radial velocity. It shows the particle moves faster closer to the boundary and the subspace enlarges faster, closer to the edge of the universe.

geometrical subspaces, it would be very difficult to apply any of those subspaces to represent the dynamic stochastic nature of our universe, since those available abstract mathematical subspaces firstly are empty and secondly, they are not temporal (t > 0) stochastic spaces. Nevertheless, Eq. (1.11) is currently available to describe a temporal (t > 0) dynamic universe, for which it shows that our universe changes with time and it is time that changes our universe, but not universe curves or changes the time-space.

Since all the postulation of our universe has treated time as independent, this includes Einstein's space-time [1], but our temporal (t > 0) universe is a time-space interdependent universe. Thus, as our universe emerged from a big bang explosion to become a physically realizable temporal (t > 0) universe, as depicted in Figure 1.8, we see that the past-time domain has no physical substance and no real time. From a mathematical standpoint we can treat the past-time domain as a four-dimensional time-space continuum in which time is an independent variable with respect to the precise past universes (i.e., space). But time independent is a virtual time that does not comply with a physical time requirement that time coexists with space or substance. And this precisely Newtonian and Einstein space-time had treated time as an independent variable with respect to space. Furthermore, our universe is an expanding subspace with time created by mass, from which we see that it is an energy conservation

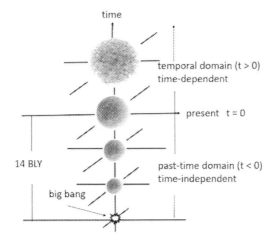

FIGURE 1.7 Shows our universe started from a big bang creation and is about 14 billion light years (BLY) old. Past-time domain (t < 0) represents some certainty, a virtual universe which has no physical substance and time. Future-time domain (t > 0) represents a physical, thely realizable uncertainty universe, and t = 0 represents an instant moment of absolute physical reality of our universe.

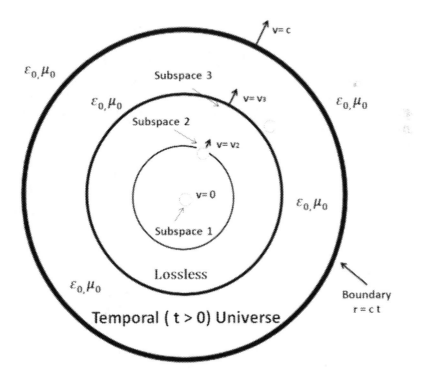

FIGURE 1.8 Shows a simplified diagram of our temporal (t > 0) universe. c is the speed of light and v is the radial velocity. In this we show that every subspace is moving radially toward the boundary of the universe, which is linearly proportional to the speed of light since light speed is the current limit.

subspace of Figure 1.7 (see Chapter 6). Which is not the same as the commonly used "zero-summed" four-dimensional spacetime continuum of Einstein [1].

Nevertheless past-time domain is a virtual domain that has no real time and physical substance in it. In other words, all that precise virtual past information is not subjected to current temporal (t > 0) constraint, although they were restricted when every past universe had been subjected before. This is precisely the reason those past-time images (i.e., information) are virtual and precisely what allows us to treat time (i.e., not real time) as an independent variable in mathematics. And this is the region that we treated time as an independent or timeless (t = 0) empty space for centuries, but not knowingly it is a virtual mathematical subspace that no longer existed. And this is also the reason all the laws, principles, and theories are time-independent or timeless (t = 0) and deterministic. Although empty space has no time and no space, mathematicians and scientists can implant virtual time and co-ordinates within an empty space (i.e., a piece of paper). From this we see where all those timeless (t = 0) laws and theories come from, since scientists are partly mathematicians.

Nonetheless, as time moves on steadily from the present absolute certainty moment into the future-time domain, we see that our universe changes with time. In this we see that time is a dependent variable such that the future moment of our universe is physically realizable. From this we see that it is a serious mistake to treat time as an independent variable within our temporal (t > 0) universe since time and space coexist. And this is the physically realizable temporal (t > 0) universe that we had not noticed for centuries, until recently. It was due the recent demands on instantaneous and simultaneous quantum computing and communication [17,18] that had prompted me to the discovery of the temporal (t > 0) universe. From this I have found that time is one of the most intriguing real variables that has no beginning and no end, and it cannot be changed or even stopped. Since time and space coexist, without space we will have no time and no space. In other words, every subspace (i.e., ΔE) cannot exist without time [i.e., $\Delta E(t)$], for which space changes with time. Although time coexists with subspace, this is by no means that if a subspace disappears time goes with it. Since subspace and energy are equivalent, from this we see that time cannot disappear with subspace, by virtue of energy conservation. Since our universe is embedded within a greater temporal (t > 0) space, we see that after our universe disappeared, its energy [i.e., $\Delta E(t)$] remains within the greater time-space, from which we see that time has no end and no beginning.

1.5   ESSENCE OF THE TEMPORAL (I.E., t > 0) UNIVERSE

In view of the preceding presentation, our universe is a time-variant system (i.e., from system theory standpoint), as in contrast with an empty space, it is a "not" a time-variant system because it has no time [i.e., timeless (t = 0)] and no space. In this we see that empty space is a virtual space "without" physical time and physical substance. Since physical reality and virtual reality are mutually exclusive, we see that temporal (t > 0) and space timelessness (t = 0) are mutually exclusive. Thus, everything that exists within our temporal (t > 0) universe cannot be empty or timeless (t = 0), which includes all the laws, principles, and theories.

Since temporal (t > 0) is a physically realizable space while timeless (t = 0) is not, any solution that comes out from an empty space paradigm is not a physically realizable solution. Yet the empty space paradigm has been used for centuries, since the dawn of science. We had not encountered any major abnormality with all those timeless (t = 0) laws, principles, and theories except a few irrational solutions in the past. This was due to the fact we had not directly confronted the temporal (t > 0) issue within the domain of our application. But not until the recent demands for instantaneous and simultaneous operation in quantum computing supremacy and as well relativistic traveling beyond and behind the pace of time. I have found recently, precisely where all those non-physically realizable principles and theories come from. It was from a simple piece of plain paper that we had normally used for analyses in the past. Inadvertently or not, we had treated the background space as an empty space and timeless (t = 0) space. Since scientists are also mathematicians, we had used this piece of innocenct paper and had never thought that it does not provide us with a real physically realizable platform. For this, scores of fictitious sciences have overwhelmed the current scientific communities that have caused a worldwide conspiracy. Whether it is truth or a fake science, that cannot be differentiated easily. And this is precisely the essence for the discovery of our temporal (t > 0) universe, from which we can differentiate a truth and a virtual science. For example, if Einstein had treated that piece of paper as a temporal (t > 0) subspace, then he would have had discovered his special theory cannot exist within our temporal (t > 0) universe. Similarly, if Schrödinger had used his quantum analyses within a temporal (t > 0) space, he would have had found that his fundamental principle of superposition failed to exist.

Nevertheless, science is a law of approximation and mathematics is an axiom of absolute certainty; using "exact" mathematics to evaluate "inexact" science "cannot" guarantee its solution exists within our temporal (i.e., t > 0) universe. One important aspect within our temporal (t > 0) universe is that one "cannot" get something from nothing; there is always a price to pay. For example, every piece of temporal (t > 0) subspace takes an amount of energy (i.e., ΔE) and a section of time (i.e., Δt) to create. In other words, time and subspace coexist or are mutually inclusive, which is the boundary condition of our temporal (t > 0) universe, of which every existence within our universe has to comply with this temporal (t > 0) condition, otherwise it cannot be implemented within our universe. Thus, we see that any postulated science has to be proven that it existed within our temporal (t > 0) universe, otherwise it is as virtual as mathematics is.

The burden of a mathematical postulation is that it needs to prove there exists a solution before searching for a solution. Although we hardly had an existent criterion in science, we need to show that a scientific hypothesis existed within our temporal (t > 0) universe, otherwise the solution would be virtual, like mathematics. For example, the superposition principle in quantum mechanics, as I have proven, it is not a physically realizable principle within our temporal (t > 0) universe, since Schrödinger's quantum mechanics is a timeless (t = 0) machine [19,20].

There is, however, an additional constraint imposed by our temporal (t > 0) universe and that is affordability. As I have shown, everything (e.g., any physical subspace) existed within our universe has a price tag, in terms of an amount of energy ΔE and a section of time Δt (i.e., ΔE, Δt), which is the necessary cost. However, to make it sufficient, we also need an amount of information ΔI or an equivalent amount of entropy ΔS (i.e., ΔE, Δt, ΔI) to make it happen. For example, the creation of a piece of simple piece of paper it will take a huge amount of energy ΔE, a section of time Δt, and an amount of information ΔI (i.e., equivalent amount of entropy ΔS) to create. But, on this planet earth only humans can make it happen. From this we see that every physical subspace (or equivalently substance) within our temporal (t > 0) universe has a price tag (i.e., ΔE, Δt, ΔS) and the question is can we afford it?

It may be interesting to know why the speed of time is a constant dependent variable. Since time coexists with substance (i.e., our temporal (t > 0) subspace), the pace of time had been long settled before the big bang explosion, otherwise the big bang creation would not have happened. From this we see that a greater temporal (t > 0) space has existed well before the creation began; otherwise the big bang explosion cannot be justified to exist (see Chapter 5). Since the current observation is limited by the speed of light, the larger temporal (t > 0) space that our universe is embedded in remains to be found. In this we see that time is a dependent-forward variable that has no beginning and no end. And this is the "time" that defines the physical reality and virtual reality. In other words, anything that coexists with time is "physically real"; otherwise, it is as "virtual" as mathematics is or an illusion like a dream. For example, are we a part of the temporal (t > 0) universe or is the universe a part of us? Yet without us, how can we tell there is a universe? The answers are we are a part of the universe that is "physically real", since we exist "with" the universe. The universe that is a part of us is an "illusion" because the universe will not go with us after we depart this universe. Similarly, we exist with time, but time will not go with us after we die. Likewise, the speed of light is dependent on the refractive index of the media (i.e., ε, μ); it is light travels with time, and it is "not" time travels with light. From this we see that light is a physical quantity and time is an invisible, "dependent", real variable. In other words, we can change the speed of light, but we "cannot" change the speed of time, although in practice we mostly assume the speed of light is constant.

1.6 PHYSICAL REALIZABLE SPACE

The physical reality of science is dependent upon the physical realizable subspace where a scientific model is embedded in; otherwise, we cannot guarantee our analysis can exist within our temporal (t > 0) space. For example, if one is designing an aircraft, the designing hypothesis should comply within the atmospheric subspace instead of within an empty space or even an underwater condition. Otherwise, the analytical solution will be fictitious, which is "not" applicable within our atmospheric environment. And this is all about science; it has to be physically realizable, otherwise it will be as virtual as mathematics. In this we see that it is a physically realizable paradigm that will take us to the promised land of physically realizable science (i.e., the truth science) but not the rigorous fancy mathematics, although science needs mathematics. In other words, it is physically realizable principle to navigate the mathematics, but not let mathematics navigate the science.

Nevertheless, as we accepted our temporal (t > 0) universe is a time and space interdependent subspace, as sketched in Figure 1.8, we see that it would be very difficult or impossible to use currently known mathematical spaces (e. g., Euclidean, Hilbert, Riemann, Banach, topological, and others) to describe our temporal (t > 0) universe, since those mathematical subspaces are virtual empty subspaces.

Yet, it is not the current dimensionality issue of those mathematical subspaces, but the nature of temporal (t > 0) of our universe or its subspaces that "prevents" us from using those mathematical subspaces for our application, unless further modifications can be found. For example, any coordinate system to describe our temporal (t > 0) space has to able to simulate non-empty temporal (t > 0) coordinates that created the space, in which time is a dependent variable with respect to the subspace.

In view of the complexity of our universe, we assume momentarily our universe is a "homogeneous lossless" temporal (t > 0) universe that compacted with "lossless" uniformly distributed refractive index substances. Since science is supposed to be approximated, every subspace within our universe has to be a temporal (t > 0) subspace; otherwise it cannot exist within our universe, from which we see that all the laws, principles, and theories have to be temporal (t > 0) laws, principles. and theories, otherwise they are as virtual as mathematics, as from a physically realizable standpoint. Since those timeless (t = 0) laws, principles, and theories were and still are the cornerstones of our science, some of them are fictious and irrational, which is like walking within a fantasy timeless (t = 0) land of science.

For example, let me show one of the commonly used atomic models, as depicted in Figure 1.9. In this we see that Bohr's atomic model [21] is situated within an empty space, depicted in Figure 1.9(a) and the other is embedded within a temporal (t > 0) space, shown in Figure 1.9(b). With reference to the temporal (t > 0) exclusive

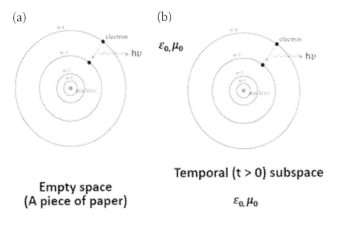

FIGURE 1.9 (a) Shows an atomic model of Bohr that is embedded within an empty space that has no time and distance. It is not a physically realizable paradigm. (b) Shows an atomic model of Bohr that is embedded within a temporal (t > 0) space. But the atomic model is not physically realizable yet since hν is not temporal (t > 0) (i.e., time limited).

principle, the paradigm of Figure 1.9(a) can immediately be disqualified. Although Figure 1.9(b) shows a step closer, the atomic model is still not yet a physically realizable model because of the idealized quantum leap energy hv, which is time unlimited, where h is Planck's constant and v is the frequency of the quantum jump. In order to make Figure 1.9(b) qualify, the paradigm has to reconfigure its quantum leap into a time and band-limited quantum wavelet of h Δv. From this, any solution, as obtained from Figure 1.9(b), will be physically realizable, which includes Schrodinger's temporal (t > 0) equation that needs to be developed.

In view of the temporal (t > 0) subspace, we see that there is a major deference from Einstein's universe, since Einstein had treated time as an independent variable with space. While our temporal (t > 0) space treats time as a "dependent" variable that coexists with the space [i.e., $\Delta E(t)$], since time coexists with space, it is by no means that time changes with the space, but it is space that changes with time. Yet space can change a section of time Δt, but it "cannot" change the pace of time. From this we see that it is not how rigorous the mathematics is, it is the "physical realizable" science that we are embracing. Although science is approximated, it should "not" be as virtual as mathematics is.

1.7   MACRO- AND MICRO-SPACE COVER-UP

Two of the important pillars in modern physics must be Einstein's relativity and Schrödinger's Quantum theory; one deals with a very large object, and the other deals with small particles. Since both of Einstein's theories and Schrödinger's mechanics were developed from an empty subspace [1,9], I will show in subsequent chapters that they are "not" physical realizable principles that can be implemented "directly" within our temporal (t > 0) universe. Since Einstein's relativistic theory and Schrödinger's quantum mechanics were derived from a virtual empty subspace platform, Einstein's special and general theories as well Schrödinger's equation are timeless (t = 0), or time independent.

Since Einstein and Schrödinger have given those fantastic "unthinkable" promises, they had led us to believe that physical behaves within a macro and a micro are different, otherwise relativistic theories and quantum mechanics cannot be reconciled. It is either inadvertent or intentionally remains to be understood. But this is the reason that I will show that particles behave within a macro- and a micro-space are basically the same regardless of the size within our temporal (t > 0) universe. From this we see that the justification for the way a particle behaves differently within a macro- and a micro-space is a major cover-up for quantum and relativistic mechanics' worldwide conspiracy. For this it allows fictitious and virtual sciences continue to spread. Although Einstein strongly opposed Schrödinger's quantum theory [10], I had have found Einstein's relativity theories had committed the same error for using the same empty space paradigm. From this we see that particles behave basically the same within a macro- and a micro-space, regardless of their size. Nevertheless, the major difference between Einstein's theory and Schrödinger's principle is that one wishes to move ahead or delay the speed of time and the other is to stop the pace of time. Yet neither move ahead nor stop the time, since our universe moves with time, but does not change the time.

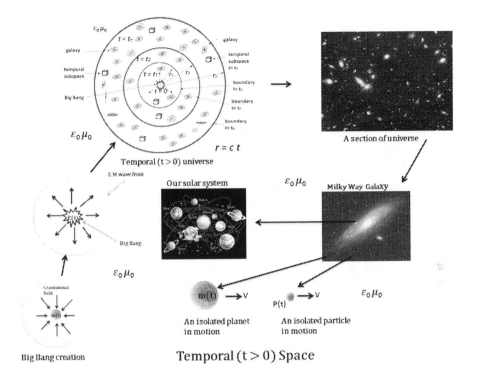

FIGURE 1.10 Shows our universe was originated by a big bang explosion from a singular temporal mass m(t) triggered by its own intensive gravitational force within a preexisted temporal (t > 0) space. In this we see that any isolated planet changes and wanders within a vast cosmological space with time, which behaves as a tiny particle in the diagram. From this we see that the behaviors within micro- and macro-space are basically the same.

As commonly agreed, a viable diagram is worth more than hundreds of equations. Once again, let me show the creation of our temporal (t > 0) universe, as summarized in Figure 1.10.

In this it shows that the origin of our temporal (t > 0) universe was started by a big bang explosion within a preexistent temporal (t > 0) space. We had shown it is the preexistent temporal (t > 0) space that allows a singularity mass m(t) to exist and grow over time, so that its induced gravitational force eventually triggers the thermo-nuclei explosion of mass m, enabling a big bang explosion. From this we see that our universe is a temporal (t > 0) stochastic universe that changes with time. And the boundary of our universe expands constantly at the speed of light, from which we see that every substance regardless of their size changes with time. And time is the only invisible real variable that runs at a constant pace, which cannot move ahead or behind its own pace or even stop. And this a physically realizable time-space that is different from Einstein's space-time continuum where he had treated time as an independent variable [1]. The fact is that the temporal (t > 0) universe is a newly discovered realizable time-space that is closer to truth. From this I would anticipate temporal (t > 0) space will eventually become one of the viable physically realizable time-space in years to come, since every law, principle, and theory developed from a temporal (t > 0) space platform will be physically realizable. In view of our universal cosmology, we can see that there are no essential differences in behavior if a particle is situated within a micro- or a macro-space, since no matter how small it is, it changes with time. And it cannot stop time as the superposition principle wishes, and also cannot slow down or move ahead time as the relativistic theory wanted.

1.8 CONE OF AMBIGUITY

Since everything within our temporal (t > 0) universe changes naturally with time, we see that within an isolated subspace, ambiguity increases naturally with time, as illustrated in Figure 1.11, which is due to the time-dependent or temporal (t > 0) stochastic dynamics process for which every subspace within our universe is temporal (t > 0) since our universe expands at the speed of light.

In this we assume a particle is situated at a precise spatial location at t = −t₃. Since the particle changes location naturally with time as our dynamic universe expands with time, we see that to predict where the particle will be located casts a shadow of possibility as time moves on. In other words, the farther in time with respect to the present moment (i.e., a section of time Δt), the more uncertain it is to predict, even under no external perturbation assumption. In other words, the higher certainty prediction is the shorter the section of Δt. In this we see that with absolute certainty prediction occurs if and only if time does not move (i.e., Δt = 0), which is equivalent to a timeless (t = 0) condition. But within our temporal (t > 0) universe, every aspect moves with time, which includes all the laws, principles, and theories

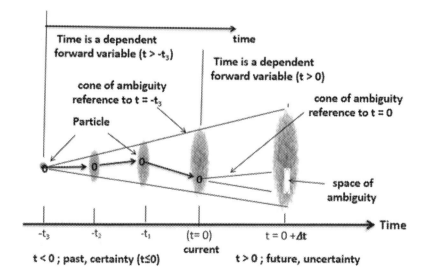

FIGURE 1.11 Shows a cone of ambiguity of our temporal universe that changes naturally with time, where the cone of uncertainty increases naturally with time farther away from absolute certainty.

and should be temporal (t > 0), or change with time. From this it includes all the classic probability distributions, and the information theory should be temporal (t > 0) since we are living within a temporal (t > 0) space whether you accept it or not. For example, information is related to entropy; since entropy changes with time, from this we see that information degrades also naturally with time (see Chapter 4).

Although the particle's past location can be retrieved based on the present moment (i.e., t = 0) of certainty, this is by no means that it is absolute deterministic since the past certainty information had to have degraded with time, as from $t = -t_1$ to t = 0 (i.e., $\Delta t_1 = |0 - t_1|$). From this we see that the further back in time, the more uncertain that past information can be retrieved, which entropy information content increases naturally with time. This is in contrast with most scientists who believe that information is preserved. And this is exactly what Boltzmann's entropy theory meant: within an isolated subspace, entropy increases naturally with time. Similarly, an isolated information content degrades with time since entropy and information are equivalent. Nevertheless, let me stress once again, within our universe every subspace has the same time and has the same speed of time. From this we have shown that it is the relativistic velocity or relativistic distance, but it is not their relativistic time as Einstein's special theory claimed (see Chapter 5).

1.9   REMARKS

In order to understand the physical realizable science, first of all we have to understand what kind of a physical subspace we are living in since it is our universe that governs science but not our science that changes the universe. For this it was my privilege to introduce how our temporal (t > 0) universe was created. Our universe is a physically realizable subspace that all physical sciences are situated within; otherwise, any hypothetical science would be as virtual as mathematics is. In other words, any scientific law, principle, theory, and hypothesis has to be proven that it existed within our temporal (t > 0) universe, otherwise it cannot actually exist within our universe. Since "time" is one of the most esoteric variables within our universe, there would be no physical substances, no space, and no life. With reference to Einstein's energy equation, I showed that our universe was created within a preexistent temporal space. From this we see that our universe is fully packed with physical substances (i.e., energy and mass). Since no physical space can be embedded in an absolute empty space, it is reasonable to assume that our universe is a subspace within a larger temporal (t > 0) space. In other words, our universe could have been one of the many universes outside our universal boundary that comes and goes like bubbles. I had showed that within out temporal (t > 0) universe, everything has a price, a section of time Δt, and an amount of energy ΔE (i.e., Δt, ΔE) and it is not free. Since ΔE and Δt coexist, it is the section of Δt that can be changed but not time since time and subspace coexist [i.e., Δt, $\Delta E(t)$], where $\Delta E(t)$ is equivalent to space or substance. Finally, I note that it is "not" how sophisticated or rigorous mathematics is, it is how mathematics relates to science. For this I stress that it should be science directs mathematics, but to not let mathematics take over the leadership of science. Otherwise, we will forever be trapped within the timeless (t = 0) wonderland of fancy science.

REFERENCES

1. A. Einstein, *Relativity, the Special and General Theory*, Crown Publishers, New York, 1961.
2. F. T. S. Yu, "Time: The Enigma of Space", *Asian Journal of Physics*, 26 (3), 143–158, 2017.
3. F.T.S. Yu, "From Relativity to Discovery of Temporal (t > 0) Universe", *Origin of Temporal (t > 0) Universe: Correcting with Relativity, Entropy, Communication and Quantum Mechanics*, Chapter 1, CRC Press, New York, 1–26, 2019. New York.
4. M. Bunge, *Causality: The Place of the Causal Principle in Modern Science*, Harvard University Press, Cambridge, 1959.
5. J. A. Fireman, A. Joshua, M. S. Turner, and D. Huterer, (2008). "Dark Energy and the Accelerating Universe". *Annual Review of Astronomy and Astrophysics*, 46 (1): 385–432, 2008.
6. J. D. Kraus, *Electro-Magnetics*, McGraw-Hill Book Company, Inc., New York, p. 370, 1953.
7. M. Bartrusiok and V. A. Rubakov, *Introduction to the Theory of the Early Universe: Hot Big Bang Theory*, World Scientific Publishing Co., Princeton, NJ, 2011.
8. F. T. S. Yu, "What is "Wrong" with Current Theoretical Physicists?", *Advances in Quantum Communication and Information*, Edited by F. Bulnes, V. N. Stavrou, O. Morozov and A. V. Bourdine, Chapter 9, pp. 123–143, IntechOpen, London, 2020.
9. E. Schrödinger, "Discussion of probability relations between separated systems". *Mathematical Proceedings of the Cambridge Philosophical Society*, 31 (4): 555–563, 1935.
10. A., Einstein, B. Podolsky and N. Rosen (*1935-05-15*). "Can Quantum-Mechanical Description of Physical Reality be Considered Complete?" (PDF). *Physical Review*, 47 (10): 777–780, 1935.
11. J. M. Knudsen, and P. Hjorth, *Elements of Newtonian Mechanics*, Springer Science & Business Media, 2012.
12. R. Zimmerman, *The Universe in a Mirror: The Saga of the Hubble Space Telescope*, Princeton Press, Princeton, NJ, 2016.
13. W. Heisenberg, "Über den anschaulichen Inhalt der quantentheoretischen Kinematik und Mechanik", *Zeitschrift für Physik*, 43, 172, 1927.
14. E. MacKinnon, "De Broglie's thesis: a critical retrospective", *American Journal of Physics*, 44: 1047–1055, 1976.
15. F.T.S. Yu, "Aspect of Particle and Wave Dynamics", *Origin of Temporal (t > 0) Universe: Correcting with Relativity, Entropy, Communication and Quantum Mechanics, Appendix*, CRC Press, New York, 145–147, 2019. New York.
16. D. F. Lawden, *The Mathematical Principles of Quantum mechanics*, Methuen & Co Ltd., London, 1967.
17. C. H. Bennett, "Quantum information and computation", *Physics Today*, 48 (10), 24–30, 1995.
18. K. Życzkowski, P. Horodecki, M. Horodecki, and R. Horodecki, "Dynamics of quantum entanglement", *Physical Review A*, 65, 1–10, 2001.
19. F. T. S. Yu, "The Fate of Schrodinger's Cat", *Asian Journal of Physics*, 28 (1), 63–70, 2019.
20. F. T. S. Yu, "Nature of Temporal (t > 0) Quantum Theory: Part II", *Quantum Mechanics*, Edited by P. Bracken, Chapter 9, p 161–188, IntechOpen, London, 2020.
21. N. Bohr, "On the Constitution of Atoms and Molecules", *Philosophical Magazine*, 26 (1), 1–23, 1913.

2 Physical Science and Virtual Mathematics

Every physical science that exists within our temporal subspace must be temporal (i.e., t > 0); otherwise, it is as virtual and fictitious science as mathematics is. The burden of a scientific hypothesis is to prove it exists within our temporal (t > 0) universe and then find the solution. In this chapter, I will show that although there is a duality between science and mathematics, any scientific postulation has to be shown it is complied with the boundary condition of our temporal (t > 0) universe, before accepting it is a physically realizable science. Otherwise, the mathematical solution cannot guarantee it is a physically realizable science. Currently, one of the important challenges must be the causality or temporal (t > 0) constraint within our universe, which is to confine an analytical solution that is temporal before accepting it as a physically realizable solution. In view of the entire fundamental laws, principles, and theories of science are mathematics, from this I have found practically all of them are timeless (t = 0) or time independent; strictly speaking, they are not physically realizable within our temporal (t > 0) universe. For example, Schrödinger quantum mechanics is a timeless (t = 0) quantum machine, from which I have found his fundamental principle of superposition does not exist within our universe. As well, Einstein's special and general theories are non-physically realizable principles since they were developed from an empty space platform. From this we see that it is not how rigorous and complex mathematics is, it is the physically realizable paradigm that determines the physical reality of science. Since empty space is a virtual mathematical space that has no time (i.e., t = 0) and no space, theoretical physicists can implant any fictitious scientific model within an empty subspace as they wish. For this, all the principles and theories are timeless (t = 0) or time independent.

Since using amazing mathematical modeling with fantastic computer simulations shows beautiful convincing animated videos, mathematical modeling and computer animation are virtual and fictitious, for which many of their simulated solutions are not physically real. What is wrong with theoretical physicists is that they used an empty mathematical subspace inadvertently for their analyses, but empty subspace is not a physically realizable subspace for any physically realizable analysis, for which I stress that it is not how rigorous, sophisticated, and complex mathematics is, it is the essence of a physically realizable mathematical paradigm. For instance, timeless (t = 0) subspace science has been used for centuries but it has produced many fictitious solutions. Yet, theoretical physicists were and still are the creators for all the fundamental laws and principles of science. But it is our responsibility to take back the physically realizable mathematics; otherwise, we will be continually trapped within a timeless (t = 0) mathematical land of non-physically realizable science.

DOI: 10.1201/9781003271505-2

2.1   SCIENCE AND MATHEMATICS DUALITY

Since science is a law of approximation, but mathematics is an axiom of absolute certainty, from which we see that using exact mathematics to evaluate approximated science cannot guarantee its solution exists within our temporal (i.e., t > 0) subspace. Since science is also a principle of logic, without logical explanation, science would be impossible to facilitate and to apply. Nonetheless, an ounce of good approximation is worth more than tons of calculations. Yet the current issue is about the burden of science, which has to be physically realizable within our temporal (t > 0) universe; otherwise, it is as virtual and fictitious as mathematics is since mathematics is not equal to science. In other words, any scientific (i.e., mathematical) solution has to comply within the boundary condition of our temporal (t > 0) space [1,2]; dimensionality and temporal (t > 0) realizability can be described by a temporal (t > 0) space equation such as:

$$\nabla \cdot S = f[r(t)], \ t > 0 \tag{2.1}$$

where $(\nabla \cdot)$ is the divergent operator, S is an energy vector (e.g., Poynting vector), and r(t) is radius of the boundary expands with time; time t is a dependent-forward variable that coexists with space. In this we see that $\nabla \cdot S$ describes a three-dimensional temporal (t > 0) space as its boundary expands at the speed of light. In other words, any scientific solution that does not comply within the boundary conditions [i.e., dimensionality and temporal (t > 0) condition] of our temporal universe, the solution is likely be as virtual as mathematics is. For example, it cannot be directly implemented within our temporal (t > 0) space. For this, we see that every subspace within our universe is a dynamic temporal (t > 0) stochastic subspace that changes with time but does not change the time. In other words, no matter how small a subspace or a particle is, it has to be temporal (t > 0); otherwise it cannot exist within our universe.

Since science needs mathematics, but mathematics does not need science, we see that science is mathematics, but mathematics is not equal to science. In other words, without mathematics, it will be very difficult to develop science. The fact is that science is a very complicated subject to deal with; it needs to prove it is physically realizable; unlike mathematics, it does not need to prove its solution is physically real. Yet any mathematical postulation needs to prove it exists as a solution first before searching for the solution. In this we found that it does not have a criterion in science to prove that its analytical solution complies within the boundary condition within our temporal (t > 0) universe, before taking it as a physically realizable solution. And it is precisely the reason why we need to understand the paradigm of a science model, is it a physically realizable model or not before pursuing all those complication mathematical analyses otherwise it would be mathematics governs the science. Since it is science that navigates mathematics, but not let mathematics dictate science, a mathematical solution is not based on the merit of how sophisticated and rigorous a solution is, it is the physically realizable science we are looking for.

Since time is space and space is time, energy and mass (or substance) are equivalent in view of Einstein's energy equation (i.e., $E = mc^2$). We see that an amount of energy ΔE is equivalent to the amount of mass Δm or subspace, in which every ΔE coexists with a section of time Δt. In other words, if it is an energy ΔE then it is a section of time Δt attached to it. But it is the section of time Δt that has been used that we cannot get back the same section (i.e., moment) of Δt, because time is a forward-dependent variable that moves at a constant pace. From this we see that we can change a section of time Δt, but we cannot change the speed of time. And this is the temporal ($t > 0$) space that we are living in. Believe it or not, this temporal ($t > 0$) space is not the space-time of Einstein [3] that I have shown in the preceding chapter you may have read.

Nevertheless, every mathematical postulation needs to prove it exists as a solution before searching for the solution. Yet in science, it seems to me it does not have a criterion as mathematics does, to prove first a hypothetical science exists within our temporal ($t > 0$) universe. Without such a criterion, fictitious and virtual science emerges as already happening in every day's scientific event. In this we have seen that practically all the laws, principles, and theories of science were timeless ($t = 0$) or time independent. Although some of those timeless ($t = 0$) laws, principles, and theories were and still are the cornerstone of our science, but there are scores of them that are fictitious or virtual, and do not exist within our temporal ($t > 0$) universe; for example, the timeless ($t = 0$) fundamental principle of Schrödinger [4] and nonphysical realizable theories of Einstein [5], which were derived from a timeless ($t = 0$) platform as I will show in Chapter 5.

Yet all the laws, principles, and theories were made to be broken, to be revised, or to be replaced. In this we have witnessed the changes from Newtonian, to Hamiltonian, to Einstein's relativity, and to Schrödinger's quantum mechanics. And this is all about science that changes with time, but does not change the time.

2.2 MATHEMATICS AND TIMELESS (T = 0) PARADIGM

For example, let me pick two of the most elegant and simple formulas in science: one is the second law of Newton [6] and the other is the energy equation of Einstein, as given:

$$F = ma \qquad (2.2)$$

$$E = mc^2 \qquad (2.3)$$

where F is the force, m is the mass, E is the energy, and c is the velocity of light. In this we see that they are point-singularity approximated equations; timeless ($t = 0$) and dimensionless. Strictly speaking, these equations cannot be directly implemented within our temporal (t > 0) universal since firstly they are not time-varying solutions and secondly they were not constrained by a temporal (t > 0) condition [i.e., every physical equation only existed within the positive time domain (i.e., t > 0)].

Since these equations provide only the amount of a physical quantity that can be traded with other physical units, but the equations show no signs of changes with time. In order to make these elegant laws temporal (t > 0), we can reconfigure them to become time-varying equations as given by:

$$F = m \ dv/dt, \quad t > 0 \tag{2.4}$$

$$\frac{\partial E}{\partial t} = c^2 \frac{\partial m}{\partial t}, \quad t > 0 \tag{2.5}$$

where t > 0 denotes that the formula is existed within the positive time (i.e., t > 0) domain. From this we see that these equations are temporal (t > 0) formulas (e.g., time-varying equations) that can be directly implemented within our temporal (t > 0) universe, even though they are still point-singularity approximated, yet science is inexact. Although these equations are physically realizable but in principle they are still not temporal equations yet, unless mass m is a temporal mass m (t) since every substance within our universe has to be temporal (i.e., by virtue of the temporal exclusion principle). This means any substance or subspace within our temporal universe is temporal (t > 0), which means that any subspace or substance changes naturally with time. From this we see that the Newtonian law as well Einstein energy equation must be temporal (t > 0) equations, as shown by, respectively:

$$F(t) = m(t) \times dv/dt, \quad t > 0 \tag{2.6}$$

$$\frac{\partial E(t)}{\partial t} = c^2 \frac{\partial m(t)}{\partial t}, \quad t > 0 \tag{2.7}$$

From this we see that every law, principle, and theory must be temporal (t > 0); otherwise they cannot be directly implemented within our temporal (t > 0) universe (i.e., by virtue of the temporal (t > 0) exclusive principle).

One of the trivial examples on the physically realizable paradigm must be the big bang creation [7] of our universe, as depicted in Figure 2.1. This is a classic example where most astrophysicists and cosmologists had assumed the big bang explosion was started from an empty paradigm, as shown in the diagram.

But, as we viewed the assumed paradigm carefully, we see that it is not a physically realizable paradigm by virtue of the temporal (t > 0) exclusive principle; substance (i.e., big bang) and emptiness are mutually exclusive.

Even though by ignoring the temporal exclusive principle, the aftermath consequence of big bang explosion would be unconceivable, as shown in Figure 2.2. In this we see that the speed divergent energy would be limited, as from electromagnetic radiation standpoint. Although it is illogical, the big bang creation has

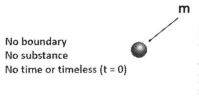

m

No boundary
No substance
No time or timeless (t = 0)

Empty space

FIGURE 2.1 Shows a big bang explosion started within an empty space paradigm as a generally accepted model by most cosmologists. m presents a point-singularity approximated mass that is assumed situated within a vast, unbounded empty space.

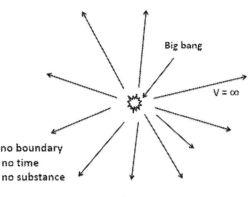

Big bang

V = ∞

no boundary
no time
no substance

Empty space

FIGURE 2.2 Shows a big bang explosion within an empty space paradigm. v is the velocity of infinity radiation within space that has no substance. Since empty space is not bounded, we cannot have a bounded universe. From this it is not a physically realizable paradigm.

been credited to the origin of time as some cosmologists assumed. However, by the Hubble space station telescope observation, we see that our universe is an expanding bounded universe changes naturally with time [8]. And this is one of the many examples shown that use a non-physically realizable paradigm, fictitious, and virtual mathematical solution emerged. From this we see that it is not how sophisticated mathematics is, it is the physical realizable paradigm determines its analytical solution is physically realizable.

Since an empty subspace paradigm is not a physically realizable model that is supposed to be used, any analytical conjecture from this paradigm is anticipated to be as virtual as mathematics is. For this, we see that it is very unlikely its solution can exist within our temporal (t > 0) universe.

Although empty space is virtual, theorical scientists can implant substances and coordinate systems they wish within an empty subspace, since theoretical scientists are also mathematicians. Since this practice has been done in the past several centuries, but not knowing the temporal (t > 0) exclusive principle, we have seen scores of timeless (t = 0) or time-independent laws, principles, and theories have developed. Let me note that timeless (t = 0) and time independent are equivalent, since any timeless (t = 0) law, principle, and theory can be used at any time without any restriction in time from mathematical standpoint. But the "physically real" time is a very restricted time, which is a dependent-forward variable moving at a constant pace and it coexists with subspace or substance. In other words, physically real time cannot exist by itself, although subspace or substance changes with time. But subspace and substance cannot change the time or even stop the time, as I had shown in Chapter 1. And this is precisely the reason that all the laws, principles, and theories we have developed are timeless ($t = 0$) or time independent. For this, we see what an empty space can do for substances, such as pretending they exist within a virtual empty space, as shown in Figure 2.3.

Within an empty space paradigm there is no time, no distance, and no boundary (i.e., strictly speaking it has no space), and we see that the implanted particle will spread instantly all over the entire unbounded timeless ($t = 0$) space from a virtual-to-virtual reality standpoint. In other words, within an empty space, we can find the particle everywhere and at every instant. And this is the virtual superposition principle, since a particle (i.e., substance) cannot coexist within a space that has no time and no substance or space.

Thus, we see that empty timeless space is a mathematical virtual space that is not an inaccessible subspace within our temporal ($t > 0$) universe, as some scientists still believe it is. And this is precisely where the simultaneous and instantaneous superposition principle of Schrödinger derived from an empty timeless ($t = 0$) space platform, but unfortunately the principle cannot exist within our temporal ($t > 0$) universe [9,10]. And this is one of the consequences that was done to quantum supremacy, which is chasing a timeless ($t = 0$) computing and within a temporal ($t > 0$) space. From this it is like searching for an angle within our time-space, but it only exists within an empty mathematical subspace.

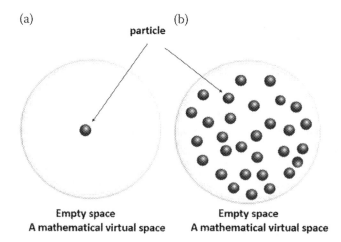

FIGURE 2.3   Shows a scenario where the same particle can be found within a virtual empty space since empty space has no time, no distance, and no boundary. Figure 2.3(a) shows mathematical physics implanted a motionless particle within a piece commonly used a scratch paper. But empty space has no time and no space, and we see that the particle spreads everywhere within an empty space, as shown in Figure 2.3(b).

We have recently found where the nonphysical realizable science comes from; it is from a piece of scratch paper that we have normally used since the dawn of theoretical science. From this we have shown it is not how rigorous mathematics is, but it is the simplicity of a physically realizable temporal (t > 0) subspace that we have used. And please tell me not the years of theoretical experience, if they were not developed from a physically realizable paradigm, it is very likely that your solution will be virtual as your mathematics.

Furthermore, how does a physically realizable paradigm affect our analyses? For example, let me show a physically realizable big bang theory, as depicted in Figure 2.4, where we see that it is a singular approximated temporal substance of m (t) embedded within a temporal (t > 0) space.

In view of this physically realizable paradigm, we see that it is a point-singularity approximated temporal (t > 0) mass m(t) that marches with the embedded temporal (t > 0) physical reality, for which the paradigm of Figure 2.4 is a physically realizable model for analyses. In view of this paradigm, time has already existed well before the assumed big bang explosion. From this we see that an induced gravitational field has a time-space to store its energy; otherwise it has no space to store within an empty space. And this is a very viable justification to trigger the big bang explosion by its own induced gravitational field over time. Otherwise, it would be very difficult to justify that the big bang explosion was ignited by itself, as commonly assumed [7]. Again, we see that a physically realizable solution depends upon a physically realizable paradigm, which has nothing to do with fantastic speculation or rigorous mathematics.

Nonetheless, science is all about approximation; otherwise science would be extremely difficult to understand and to facilitate, even with the help singularity approximation. Yet it must be due to the physically realizable subspace that safeguards a realizable big bang theory, as shown in Figure 2.4. From this we see that time is not only real, but also a dependent variable that exists with our temporal (t > 0) universe well before the big bang; in other words, time has no beginning and has no end. Yet there was a question that has been raised: what has time done to Newton's law and Einstein's energy equation? I am sure these two formulas were derived from an inadvertently empty timeless (t = 0) space, which included all the laws, principles, and theories that we are using today were developed from the same empty space platform. From this I will show that it is the temporal (t > 0) or time-dependency theory, principle, and law that determines the

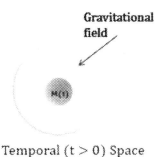

Gravitational field

M(t)

Temporal (t > 0) Space

FIGURE 2.4 Shows a physically realizable big bang paradigm, which is a singularly approximated temporal (t > 0) mass m(t) situated within a temporal (t > 0) space. In this it shows m(t) changes with time.

physical realizability. Instead, timeless (t = 0) or time-independency laws, principles, and theories predict the future with certainty. But science is approximated, which changes naturally with time.

Since science was basically started from classical Newtonian mechanics to statistical, Hamiltonian, to modern relativistic and quantum mechanics, for this we see that the Newtonian second law was developed on a piece of paper, as depicted in Figure 2.5, where the background is treated as a timeless (t = 0) empty pace. From this, all the laws, principles, and theories are timeless (t = 0) or time independent since all of them were derived from the same empty subspace paradigm. And this is precisely the reason that all the laws, principles, and theories are deterministic, although science is supposed to be inexact or nondeterministic. Since timeless (t = 0) or time independence is a semantic term, which means that law, principle, and theory does not change naturally with time or temporal (t > 0), they are time-independent principles.

The Newtonian second law is a typical example of a timeless (t = 0) or time-independent law, as depicted in Figure 2.5. Although it is not A physically realizable paradigm, we have inadvertently used this timeless (t = 0) model for centuries.

Alternatively, if Newton's second law is implemented within a temporal (t > 0) subspace, as depicted in Figure 2.6, it is a physically realizable paradigm since m(t) is a temporal (t > 0) mass (i.e., coexist with time); otherwise it is not A physically realizable model because A timeless (t = 0) particle cannot exist within our temporal (t > 0) universe.

In view of Newtonian paradigm as embedded within a temporal (t > 0) subspace, we see that it is basically a physically realizable model, but it is still not sufficient since the law is still not a time-variable equation and also not a temporal (t > 0) solution. However, if the Newtonian law is reconfigured as given by:

$$F(t) = m(t)dv/dt, \quad t > 0 \qquad (2.8)$$

where t > 0 denotes that equation exists only within the positive time (t > 0) domain. From this we see that Eq. (2.8) is a temporal (t > 0) second law of Newton, which is a physically realizable law that can be implemented within our temporal (t > 0) universe, since the equation changes with time.

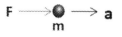

FIGURE 2.5   Shows Newtonian second law embedded within a timeless (t = 0) platform. F is the force, m is a timeless (t = 0) mass, and a is acceleration.

A piece of paper
Timeless (t=0) Subspace

FIGURE 2.6   Shows Newtonian law as implemented within a temporal (t > 0) subspace. F is the force, m(t) is a temporal (t > 0) mass, and a is the acceleration.

Temporal (t>0) Subspace

Although there are uncountable examples that can show that the essence of realizable paradigm affects the physical solution, I will use one of the most revolutionary ideas of time, such as the relativistic-time of Einstein, as depicted in Figure 2.7:

$$\Delta t' = \frac{\Delta t}{\sqrt{1 - v^2/c^2}}$$

where Δt is a section of time that has been used by a moving coordinate system (i.e., x', y', z') and $\Delta t'$ is a time dilation of the moving coordinate space (i.e., x', y', z') with respect to the stationed coordinate system (i.e., x, y, z). Since Einstein's special theory was developed within an empty timeless (t = 0) subspace that is not a physically realizable paradigm, we see that both coordinate systems' back subspace has no distance and no time. Yet Einstein had derived one of the most intriguing theories that he might not know himself, that the background of that piece of paper represents an empty timeless (t = 0) space. Otherwise, he might not have developed his theory. Aside from the non-physical realizable issue, his special theory is directionally independent. Because within an empty space there is no time and no direction, strictly speaking it has no space. Aside from the directional independent, both Einstein's theories (i.e., special, and general theory) are non-physically realizable theories within our temporal (t > 0) universe, as will be shown in Chapter 5. From this we see how important a physically realizable paradigm is. From this it shows us that it is not the complexity of mathematics, but it is the physically realizable paradigm that produces a physically realizable solution.

Yet, two of the most important pillars in modern physics must be Einstein's relativistic mechanics and Schrödinger quantum mechanics. But both of them committed the same mistake for using a non-physically realizable empty platform for their theories (see Appendix D, E). Since Schrodinger's quantum theory is a legacy of Hamiltonian mechanics, which is a timeless (t = 0) or time-independent classical mechanic, Schrödinger's machine is also time independent or a timeless (t = 0) quantum mechanics [4,9,10]. For example, Schrodinger's quantum machine was derived based on Bohr's atomic model, as depicted in Figure 2.8. This is a typical Bohr's atomic model embedded within an empty timeless (t = 0) space (e.g., a piece of paper), and apparently it is not a physically realizable paradigm.

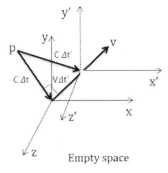

FIGURE 2.7 Shows Einstein's special theory of relativity was developed from an empty subspace (i.e., on a piece of paper) platform, which is not a physically realizable paradigm. Nevertheless, with reference to the right-triangular formulation in the diagram, Einstein's special theory of relativity can be derived from the Pythagorean theorem as given by [3].

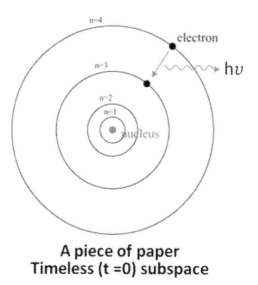

A piece of paper
Timeless (t =0) subspace

FIGURE 2.8 A classic Bohr's atomic model embedded within an empty subspace.

In view of this classic Bohr's atomic model, it seems to me the quantum state radiation hv ignored a couple of physically realizable facts. For example, substance cannot coexist within an empty space, quantum jump is not a time and band-limited physical reality, and more serious is that quantum state radiation cannot actually radiate within an empty space. From this we see that Schrodinger's equation is timeless (t = 0) or time independent, as well as his fundamental principle of superposition is timeless (t = 0) and others (see Appendix D).

Since Schrödinger's fundamental principle of superposition was based on the Hamiltonian–Bohr timeless (t = 0) model to develop his Schrödinger equation for which his equation is time independent, that cannot be directly implemented within our temporal (t > 0) universe [11]. Aside from certain doubts on the principles in the past, this nonexistent fundamental principle has aroused a worldwide quantum excitement for instantaneously and simultaneously quantum computing [12] as well quantum entanglement [13]. This depleted a huge attention for searching for a quantum supremacy, which does not exist within our temporal (t > 0) universe, but only existed within a timeless (t = 0) mathematical virtual space. From this we see that what mathematics does to science if we let mathematics controls our science instead of letting science lead the mathematics (see Appendix E).

Although a score of time-independent or timeless (t = 0) laws and theories are the facts of classics, laws, principles, and theories were made to be broken, be revised, or be eradicated. And this is all about mathematics and science for which science must be physically realizable and mathematics does not. In other words, it is science that governs mathematics but do not let mathematics dictate our science. For example, I have seen an article published by a group world-renowned mathematicians and physicists using fancy mathematics in searching for a ten-dimensional subspace within our universe [14]. My question is that is their newly founded ten-dimensional subspace a temporal $(t > 0)$ subspace; otherwise, how they can justify that their ten-dimensional subspace exists within our temporal $(t > 0)$ universe?

2.3   CLASSICAL AND POINT-SINGULARITY SCIENCE

Virtually all the fundamental laws and principles of science are mathematically simple and elegant, which makes complicated science easy to understand and to facilitate. Since science relies on mathematics, I have seen mathematics overtaken by the leadership of science for decades. From this I have found that science as been hijacked by mathematics, since theoretical scientists are more interested in the complexity and rigorous analyses that have abandoned the essence of the physical reality of science. From this I have found that we have gradually lost our original independent logical thinking and willingness to accept the complexity analyses that have been regarded as serious solutions, yet science is supposed to be simply approximated. For example, theoretical scientists underscored the simplicity of science by regarding it as not a serious solution and have overrated mathematics' complexity over the physical reality of science. Yet, fancy mathematics were not developed by the scientists, but by pure mathematicians. For this, it is important for scientists to take back the leadership of physically realizability of science instead of perpetually being trapped within the timeless $(t = 0)$ land of virtual mathematics, although science needs mathematics. Thus, as I see it, it is a serious mistake to let mathematics dictate our science, instead of letting science govern mathematics.

The fact is that our predecessors precisely knew all those simple and elegant laws and principles were approximated but implicitly with statistical inclined. Otherwise, those laws and principles cannot be presented in simple and elegant mathematical forms, which are easy to understand and to facilitate. As a general rule scientist uses past deterministic knowledges to predict new law, principle, and theory but the predicted result is deterministic, yet science is supposed to be nondeterministic or approximated. In fact, most viable laws, principles, and theories were presented by simple and elegant equations. For this, we see that rigorous mathematics is not equal to physically real science, although science is dependent on mathematics. As a scientist, partly it is my obligation to let us know that our science has been hijacked by virtual mathematics since a score of illogical and fictitious sciences has emerged. For example, the big bang creation exploded within an empty space, nonexistent superposition principle applied to quantum computing supremacy, nonphysically realizable relativity for time traveling, and many others. Although those principles and theories were basically mathematically correct, unfortunately they are physically realizably wrong. Personally, I felt that we should not let mathematics take the helm of our valuable science, otherwise what would be our role in science. For this, I have shown that if a hypothetical principle is not physically realizable, it is very likely that the postulated principle is a nonphysically realizable theory that is as virtual as mathematics is. For example, Schrödinger's fundamental principle and Einstein's relativity theory; from this we see that it is not how fancy the mathematics are, it is how we wisely use the mathematics.

Nevertheless, let me show a set of simple and elegant timeless (t = 0) or time-independent laws, principles, and theories:

$$\nabla \cdot E = \rho/\varepsilon_0 \tag{2.9}$$

$$\nabla \cdot B = 0 \tag{2.10}$$

$$\nabla \times E = -\partial B/\partial t \tag{2.11}$$

$$\nabla \times B = \mu_0 \ (J + \varepsilon_{\partial E/\partial t}) \tag{2.12}$$

$$F = ma \tag{2.13}$$

$$E = h\upsilon \tag{2.14}$$

$$P = h/\lambda \tag{2.15}$$

$$E = mc^2 \tag{2.16}$$

$$\Delta t' = \frac{\Delta t}{\sqrt{1 - v^2/c^2}} \tag{2.17}$$

$$\frac{\partial^2 \psi}{\partial x^2} + \frac{8\pi^2 m}{h^2}(E - V)\psi = 0 \tag{2.18}$$

$$\mathcal{H} = -[h^2/(8\pi^2 m)]\nabla^2 + V \tag{2.19}$$

Since all of these simple and elegant laws, principles, and theories are point-singularity approximated, without singularity approximation we would not have all those laws. It is also because of the simplicity in mathematical forms that their physical significances are easy to understand, to learn, and to facilitate.

Although a point-singularity approximated equation deviates away from its physical reality somewhat, but science is approximated. From this we see that without point-singularity approximation, it would be extremely difficult to present an elegant mathematical equation, which is easy to understand and to facilitate. For example, let us take a quantum state energy representation as given by E = hυ, where h is Planck's constant and υ is the frequency. In this, υ is a singularity approximated frequency, with no bandwidth and time unlimited. From this we see that is a non-physically realizable quantum state energy and should not be used in the first place since every quantum leap energy has to be band and time limited.

Another trivial example is Newtonian's second law F = ma; it has no dimension and m is point-singularity approximated and it is not temporal (t > 0), which means that it changes with time. But from a physical reality standpoint, every mass has a finite size, no matter how small it is. In fact, every mass or substance must be a temporal ($t > 0$) (i.e., has time) substance; otherwise it cannot exist within our temporal ($t > 0$) universe. In this we see that using a singularly approximated equation to analyze a physical problem it is very likely its solution will be point-singularity approximated [15]. But a more contingent demand is that every substance has to be a temporal ($t > 0$) substance. From this we see that all the timeless ($t = 0$) laws, principles, and theories are not physically realizable principles from a strictly speaking physically realizable condition.

2.4   WHAT MATHEMATICS HAS DONE TO SCIENCE

Since science has moved closer to a real-time regime, such as applied to information transmission and digital data computing, the demand for instantaneous and simultaneous response is needed. In this we see that the nature of temporal ($t > 0$) limit cannot be ignored. From this I have found that all the laws, principles, and theories are timeless ($t = 0$) or time independent; strictly speaking they cannot be directly implemented within our temporal ($t > 0$) subspace, which includes all the equations that I had shown in preceding equations [e.g., Eq. (2.9) to Eq. (2.19)]. Although the foundation of our science was developed from a timeless ($t = 0$) platform, a number of laws, principles, and theories are fictitious and virtual as mathematics is what I will discuss, such as Schrodinger's fundamental principle of superposition, Einstein's special and general theory of relativity, and possibly others. For this, I am taking the benefit to show what a non-physically realizable timeless ($t = 0$) paradigm can do to science.

Firstly, let me show an example of what has been done by a non-physically realizable empty subspace platform to quantum computing. Schrödinger's fundamental principle of superposition promised us a simultaneous and instantaneous superposition principle that motivated scores of worldwide computer scientists searching for quantum computing supremacy to overtake the current digital computers. Firstly, let me remind you that within our temporal ($t > 0$) universe there is always a price to pay; a section of time Δt and an amount of energy ΔE (i.e., Δt, ΔE). For example, every bit of information takes a section of time Δt and an amount of energy ΔE to transmit and to compute. For instance, if we want a bit information to be quickly transmitted, we need the smallest possible section of time (i.e., $\Delta t \to 0$). Of this we also need a very large amount of energy (i.e., $\Delta E \to \infty$) within that section of time Δt to spend (i.e., $\Delta t \Delta E = h$). Since $\Delta t \to 0$ but not equal to zero (i.e., $\Delta t = 0$), we see that it is impossible to transmit a bit of information instantaneously without a section of time to spend, since time signals need a section of Δt to present a signal, no matter how small a section Δt it is. Nevertheless, it is possible to transmit a bit or fraction of a bit of information within an empty space since empty space has no time and no distance. Then we see that a time-signal without a section of time has no signal to transmit. Since empty space cannot exist within our temporal ($t > 0$) universe, then how can we transmit a time-signal within our temporal ($t > 0$) space instantaneously without paying another section of time $\Delta t'$ to transmit? For example, we are transmitting a rectangular pulse time signal of Δt to a satellite 1 second away (i.e., $\Delta t''$) for which $\Delta t'' > \Delta t$, then we see that it will take a section of time $\Delta t' = \Delta t + \Delta t''$ to complete the transmission, which is not instantaneous, and it is not free. Even we assume the space between the transmitter and the satellite is empty, but a transmitted pulse signal still needs Δt to complete the transmission although the empty subspace cannot exist within our universe.

Empty space is not an inaccessible space as some scientists assumed it is. Empty space is a nonexistent space within our temporal (t > 0) universe by virtue of the temporal (t > 0) exclusive principle (see Appendix A). And this is one of several pieces of evidence that a non-physically realizable principle is accepted as a viable theory for quantum computing supremacy. A heavy price has already been paid that a huge amount of revenue has been allocated for a non-existent superposition principle, inadvertently or not.

Now let me simulate a virtual empty space paradigm when a pair of temporal (t > 0) wavelets are submerging into a timeless (t = 0) space paradigm, although it is not a physically realizable example. For simplicity, we assume two separated time-limited quantum state wavelets (i.e., equivalently two photonic particles) (see Appendix D) are plunging into a timeless (t = 0) space, as depicted in Figure 2.9 (a). Since distance is time within a temporal space, we have the following relationship:

$$d = c \ t \tag{2.20}$$

where c is the velocity of light and t is time. But within an empty space it has no time and no distance; in other words, every point-singularity substance will be superimposed or collapse all together at time t = 0. At the same moment, all the substances (or particles) will simultaneously exist or at the same moment of time t = 0 since empty space has no space or no distance. From this we see that empty space is a virtual mathematical timeless (t = 0) space that we have been inadvertently using it for centuries. Since empty space is a space that has no time and no distance, does it behave like Alice's wonderland? From this we see that only mathematicians and theoretical scientists can do it, since theoretical scientists are also mathematicians.

Let us show what empty timeless (t = 0) space can do to a set of wavelets, as presented by a system analog diagram shown in Figure 2.9. In this we see the output response from a timeless (t = 0) subspace collapses at t = 0 [i.e., Figure 2.9 (c)], which has lost all the wavelets' physical personalities time and frequency. Now, if we take this output response implanted within a temporal (t > 0) subspace, which is the quantum computing scientists anticipated for, but the output response from the temporal (t > 0) space does not show any of the simultaneous and instantaneous properties, as can be

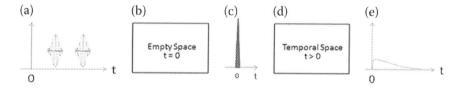

FIGURE 2.9 Shows a system simulation for timeless (t = 0) superposition within a temporal (t > 0) space. (a) Shows a set of time-limited wavelets (or particles) plunge into a timeless space of (b); (c) shows the output response collapse at t = 0; (d) shows a temporal space representation; and (e) shows the corresponding response within a temporal space, in which we see that output loses all the original input personality.

seen from Figure 2.9 (e). From this we see that in spite of the non-physically realizable issue, the system analog presentation shows that the output response is not the fundamental principle of superposition that quantum computing and communication scientists are anticipating. The answer is that the timeless ($t = 0$) superposition principle cannot exist within our temporal ($t > 0$) universe since Schrödinger's quantum mechanics was developed within an empty space platform (see Appendix E).

2.5 PHYSICALLY REALIZABLE SCIENCE

As we accepted the creation of our temporal ($t > 0$) universe, dimensionality is one of the issues that in principle it allows multi-dimensional subspace within our three-dimensional temporal ($t > 0$) space. But any higher dimensional subspace has to be temporal ($t > 0$). In other words, any higher dimensional subspace firstly cannot be empty and secondly subspace cannot be empty since time and subspace coexist. From this I have seen that mathematical scientists have evaluated a ten-dimensional subspace within our universe [14], but my question is if their multi-dimensional subspace is a physically realizable subspace; otherwise it is a mathematical virtual space that does not exist within our temporal ($t > 0$) universe.

Let me further stress that dimensionality within a temporal ($t > 0$) subspace may not necessarily have to be a Euclidean-type subspace as commonly assumed, but it is subspace and time coexisted issue that is currently limited, such that space changes naturally with time but space does not change the time or even stop the time. In other words, every subspace within our universe is a temporal ($t > 0$) space compacted with substances. Therefore, any one-dimensional or two-dimensional subspaces or even a point-singularity cannot actually exist within our universe; since every physical substance has mass and size that coexist with time. In this we see that every subspace is a dynamic time-interdependent stochastic subspace [1,2]. Which cannot be simply described by any currently known geometrical mathematical spaces, since temporal ($t > 0$) space is not an empty space as most of the virtual mathematical spaces have assumed. In other words, it is the dynamic subspace that changes naturally with time, as may be seen from our temporal ($t > 0$) universe standpoint, as depicted in Figure 2.10. In this we see that every subspace within our universe is a temporal ($t > 0$) subspace that cannot be simply described by a simple mathematical equation.

Since every subspace within our universe takes an amount of energy ΔE and a section of time Δt to create, we see that a unit of (ΔE, Δt) is a necessary cost but not sufficient to create a temporal ($t > 0$) subspace. For example, to create a drop of water, it needs an amount of ΔE and a section of Δt to make it a necessary cost to pay, but it is the amount of information ΔI or equivalently an amount of entropy ΔS to make it sufficient.

In view of the stochastic nature of our universe, it would be very difficult to find an available mathematical subspace that can be used to simulate a stochastic temporal ($t > 0$) subspace, since every temporal ($t > 0$) subspace is compacted with substances that change naturally with time, and time is a dependent-forward variable. For example, topological space is one of the abstract mathematical subspaces that scientists have hinted at for the time-space application, as depicted in

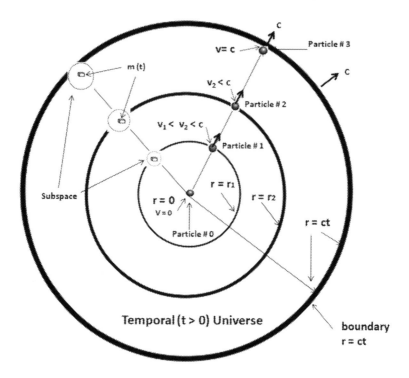

FIGURE 2.10 Shows that every subspace within our universe is a dynamic stochastic subspace that changes naturally with time. In this we see that its boundary expands at the speed of light.

Figure 2.11. In this we see that a set topological equivalent subspaces change (i.e., curves by human) with time. Since those subspaces will eventually become the certainty events (i.e., information without substance and no time), we see that those past topological equivalent spaces are independent with time (i.e., virtual times) from a mathematical standpoint. Then, if we use those past events to predict a future topological space, then the solution is anticipated to be deterministic or certaint. But a future prediction is uncertain or nondeterministic since a future prediction is supposed to be uncertain or non-deterministic. We see that space changes with time, due to human intervention. From this we see that it is not correct to use a timeless (t = 0) or time-independent subspace paradigm to predict the future universe or any future subspace that includes the future science, since science is supposed to be approximated and nondeterministic that changes with time. From this we see that it is hard to convince us that matter can curve time-space since the temporal (t > 0) space cannot change time. Again, we see that a physically realizable paradigm determines the physically realizable solution, which is not entirely dependent on the complexity and severity of mathematics, but it is dependent more on a physically realizability paradigm. In other words, mathematics is a tool that enables us to obtain a viable solution but not the deciding factor that its solution is physical realizable, which is more dependent on a physically realizable paradigm that has been used.

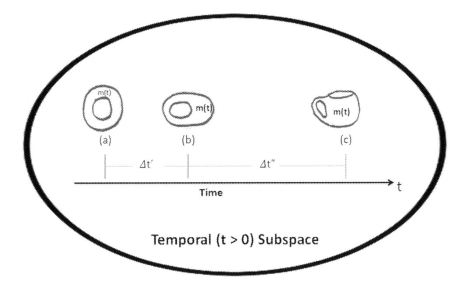

FIGURE 2.11 Shows a set of topological subspaces that change with time within a temporal (t > 0) space. Once the change is done, then those equivalent topological spaces have no physical substance and no time in it, since they represent a set of past certainty consequences similar to a backward video clip. From a physical standpoint, these subspaces no longer exist in physical realities.

Since a physically realizable subspace paradigm is crucial for scientific reality analyses, let me epitomize a time-space diagram, as depicted in Figure 2.12, to describe the nature of temporal (t > 0) space that changes naturally with time. In this we see that the present moment t = 0 divides the past virtual reality (e.g., events without real substance and no time) and future physically realizable realities (i.e., will have real physical substances and time). From this we see that within the past-time domain (i.e., t < 0) time can be treated as an independent variable from the mathematical standpoint. Similar to a playing a back and forth video clip, we can see certainty images but no real physical substances in it. While within a future-time domain (i.e., t > 0), time is a dependent variable coexisting with space, and the situation is rather different; the physically realizable subspace has physical substances that coexist with time, where time is a forward-dependent variable. And the present moment t = 0 is the only absolute physical reality, in which every subspace or substance at this moment of t = 0 absolutely exists but disappears instantly at Δt → 0 to becoming a brand-new absolute subspace and so forth. And this must be where I can briefly describe the nature of a temporal (t > 0) space. From this we see that science is not supposed to be exact but approximated. Therefore, rigorous, complex mathematics does not guarantee the physical realizability of science unless its mathematical paradigm is physically realizable.

Since the past-time domain has no time and substance in it, they were certainty subspaces located at precise past-time moments, but no time and no physical

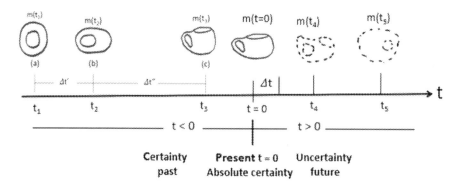

FIGURE 2.12 Past-time domain (t < 0) shows a set of topological equivalent virtual subspaces that have no physical substances and time; a future-time domain (t > 0) represents a couple physically realizable uncertain topological equivalent subspaces, and present moment t = 0 shows a n absolute physically realized topological equivalent subspace, which has a physical substance [i.e., m(t = 0)] with time, which moves forward to become a newer present moment t = 0 + Δt [i.e., m(t = 0 + Δt], where Δt is approximately equal to zero (i.e., t ≈ 0).

substances. From this we see that all the past-equivalent subspaces from t_1 to t_3 are predictable subspaces if we walk back in time. But this is the past-time domain (i.e., t < 0) where we can never be able to get back to the past subspaces, since they are virtual (i.e., memories) without physical substance and time. From this the present moment (i.e., t = 0) moves forward incrementally (i.e., Δt → 0) to a new absolute certainty at t = 0 + Δt. But it is the section of time Δt that has gone by and we cannot get it back, since subspace changes with time.

However, if we can walk back into the past-time domain (t < 0), which is a virtual space without physical substance and no time, it is the past subspace changes with time since all the past-time subspaces at $-t_1$, $-t_2$, ... are deterministic subspaces that existed at a precise neg-time domain (i.e., t < 0). From this we see that the past-time domain is time independent since the past-time domain has no time and physical substance. And this is precisely where Einstein developed his general theory for which his general theory is deterministic, but the prediction is supposed to be non-deterministic or uncertain. In other words, the derivation of general theory was developed from an empty space platform, which is a non-physically realizable paradigm. From this we see a physically realizable theory derived from a physically realizable paradigm, but not from the severity of mathematical complexity.

2.6 MATHEMATICS SUPREMACY AND CONSPIRACY

Nevertheless, a fundamental question is we have buried ourselves within a complicated timeless (t = 0) science and thrived. Or are we willingly to change to a temporal (t > 0) science? We have gradually lost our independent logical thinking and have opted to accept the approval of the others. For this, we have been more likely to accept serious complicated mathematical solutions instead of a simple physically realizable theory and principle, which had been regarded by most theoretical physicists as not a serious solution. In view of mathematical supremacy, we have been overlooking the physical realizability of science, and scores of non-physical realizable sciences emerged worldwide from renowned scientists, playing their mathematical games instead of the physical realizability of science, either inadvertently or intentionally, I will never know. Nevertheless, two of the most important pillars of modern physics must be Einstein's relativity theories and Schrödinger's quantum mechanics; one is dealing with very large objects and the other is manipulating very small particles. Because of qubit computing, which potentially can overtake the current digital computer, I would take this example as my first topic to note that qubit information only exists within an empty timeless (t = 0) space (see Chapter 4).

But before I get started, it is important to show that a non-realizable paradigm had led us to unthinkable consequences in science. Firstly, it must be the powerful mathematics used in quantum mechanics that let us to believe the principle does exist within our time-space. But I have found that Schrödinger's quantum mechanics is timeless (t = 0), virtual, and fictitious, for which it was the empty space paradigm that hijacked the physical reality of the quantum theory. For example, let me show what a timeless (t = 0) subspace does to particles, as illustrated in Figure 2.13; we see that two particles are situated within an empty space.

Since empty pace has no time and no space, these two particles are instantly superimposing together all over the entire unbonded space. Thus, we see that how easily virtual mathematics can fool the physical reality. From this the superposition principle has created a worldwide qubit quantum conspiracy. And this is one of

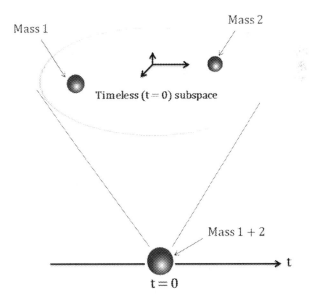

FIGURE 2.13 Shows a set isolated particles that exist within an empty timeless (t = 0) space. Since empty space has no time and no distance or space, we see that particles are superimposed together and collapsed at time t = 0. But it not a physically realizable paradigm by virtue of the temporal exclusive principle.

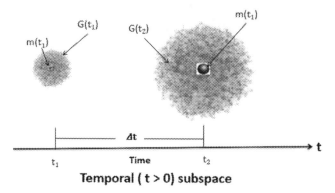

Temporal (t > 0) subspace

FIGURE 2.14   Shows a temporal (t > 0) mass m(t) is situated with a temporal space at time t = t_1. Since a substance changes with time within a temporal (t > 0) space, we see that mass m(t) grows, as well as the gravitational field G(t) grows with time at t = t_2.

many typical examples where mathematics has overshadowed the physical realizability of science. Inadvertently or not, we have already been trapped within a virtual wonderland of mathematics.

The other example must be Einstein's general theory since the theory predicts matter curves space-time. By means of a strong gravitational field, we assume a temporal mass m(t) is situated within a temporal (t > 0) space, shown in Figure 2.14. Since every temporal (t > 0) subspace is compacted with substances, we anticipate that an induced gravitation field G(t) increases with time as mass m(t) increases. For this, it takes a section of time Δt to increase the strength of its induced gravity. And it is a section of Δt needed for induced gravitational field G(t) to grow, but it is not the gravitational field G(t) that curves the speed of time. This example shows that it is not the rigorous mathematics that decides the physically realizable solution, but it is the physically realizable paradigm that determines the physically realizable science.

Nevertheless, over centuries we have used an empty space platform for analyses. Although it produced scores of viable and elegant laws and theories, at the same time it has given us a number of virtual, fictitious, and irrational principles and theories. If we continuity allow mathematics to taking over our science, we will have more fictitious science emerge that will overwhelm our science. For example, a couple of renowned mathematicians claimed they have discovered a ten-dimensional subspace within our universe [14]. Yet, my question is if their subspace is a temporal (t > 0) subspace. Otherwise, it might be as virtual as mathematics is since mathematics is not equal to science.

2.7   WHAT IS A PHYSICALLY REALIZABLE SOLUTION?

Science changes with time, and physical realizability also changes with the changes of science. Since every problem has a multi-solution, a non-existent solution is not a physically realizable solution. For example, a timeless (t = 0) or time-independent solution is not a physically realizable solution. From this we see that the difference between physical science and mathematics must be the physically realizability divides mathematics and science. But it is an irony that science needs mathematics, but mathematics does not need science.

Since the apparent division between science and mathematics is apparent, is it possible to reconfigure a timeless (t = 0) or time-independent solution to become physically realizable? The answer is possible but may not always be. For example, iet us take Schrödinger's equation as given by:

$$\frac{\partial^2 \psi}{\partial x^2} + \frac{8\pi^2 m}{h^2}(E - V)\psi = 0 \qquad (2.21)$$

where ψ is the Schrödinger wave function, m is the mass, E is the energy, V is potential energy, and h is the Planck's constant. Since Eq. (2.21) is a time-independent equation, which means that any solution comes out from this equation will be certainly time-independent or timeless (t = 0), since timeless and time-independent are equivalent.

Nevertheless, if we subjected this question by the temporal (t > 0) condition as given by:

$$\frac{\partial^2 \psi}{\partial x^2} + \frac{8\pi^2 m}{h^2}(E - V)\psi = 0, \quad t > 0 \qquad (2.22)$$

Then any solution that comes out from this equation has to comply with the temporal (t > 0) condition that guarantees its solution is temporal (t > 0). For example, I will show in subsequent chapters a non-physical wave function obtained from the traditional Schrödinger equation can be reconfigured to become temporal (t > 0), which is physically realizable. From this temporal (t > 0) wave function, it is trivial to see that Schrödinger's fundamental principle of superposition cannot be supported within our time-space.

Similarly, can we reconfigure Einstein's special theory of relativity? The answer is that we cannot. For example, if we submerge Einstein's hypothesis within a temporal (t > 0) space platform, you can easily find out that the special theory cannot produce a section of time gain. From this we see that the special theory is not a temporal (t > 0) theory that existed within our time-space.

Furthermore, as for Einstein's general theory of relativity, we can condition a temporal (t > 0) constraint at the end of the equation as given by:

$$G_{\mu\nu} + \Lambda g_{\mu\nu} = (8\pi G/c^4)\, T_{u\nu}, \quad t > 0 \qquad (2.23)$$

where t > 0 denotes that the equation is subjected by a temporal constraint, $G_{\mu\nu}$ is the Einstein tensor, $g_{\mu\nu}$ is the metric tensor, $T_{u\nu}$ is the stress-energy tensor, Λ is the cosmological constant, G is the Newtonian constant of gravitation, and c is the speed of light in a vacuum. For this, I believe that any solution that comes out from this equation would be temporal (t > 0) in principle it can be used within our time-space.

From these examples we see that Schrödinger's principle is primarily to stop time and Einstein's theory is to save a section of time, but neither stops nor saves time, since we change with time, but we cannot change or stop time. Nevertheless, these examples show that it is not the severity of mathematics, but it is the physical realizability of science that we embraced, although science needs mathematics.

2.8   PHYSICAL AND MATHEMATICAL INTERPRETATION

Since laws and principles were presented by mathematical equations, yet the interpretation could be different as from mathematical or as from physical standpoint. For example, let us take Einstein's energy equation as given by:

$$E \approx mc^2 \tag{2.24}$$

Or equivalently in partial differential form as given by:

$$\frac{\partial E(t)}{\partial t} \approx c^2 \frac{\partial m(t)}{\partial t} = \nabla \cdot S, \quad t > 0 \tag{2.25}$$

where $t > 0$ denotes equation is subjected by temporal ($t > 0$) constraint, E is the amount of created energy, m is the mass, c is speed of light, c ($\nabla \cdot$) represents a divergent operator, and S is an energy vector. As from physical significant standpoint, these equations interpret that as an amount of mass m annihilates, it converts an equivalent amount of energy left behind the subspace. From which we see that, our universe is a dynamic energy conservation expanding subspace (see Appendix E, F).

Nevertheless, if we presented Einstein's energy equations as given by respectively:

$$E - mc^2 = 0 \tag{2.26}$$

$$\frac{\partial E(t)}{\partial t} - c^2 \frac{\partial m(t)}{\partial t} = 0, \quad t > 0 \tag{2.27}$$

we see that these two equations are essentially identical as preceding equations. However from physical standpoint, this set of equations interpret as zero-summed equations, as most physicists do. And this precisely the reason that our universe had been wrongly interpretated as zero-summed energy universe. Since our universe was created by $E = mc^2$, from we see that our universe is an energy conserved subspace (see Chapter 6), instead of a zero-summed universe as most theorical physicists believed (e.g., the four-dimensional space-time continuum).

With reference to the famous Dirac equation as given by [16]:

$$(i\partial\!\!\!/-m)\psi = 0 \tag{2.28}$$

where $\partial\!\!\!/$ is a Feynman slash notation (which is not an essence factor for our current issue), m is the mass of a particle, and ψ is the Schrödinger wave function. In view Dirac equation we see that, it is essentially a zero-summed mass equation (i.e., or energy equation since mass and energy are equivalent). And this is precisely the equation that Dirac had hypnotized the existence of anti-particle within our universe. Since anti-particle traveling backwardly in time (i.e., $t < 0$), we see that it is impossible for anti-particle to exist within our temporal ($t > 0$) universe, since our universe changes forwardly with time (i.e., $t > 0$).

Nevertheless, if Dirac theory is presented by a non-zero summed equation as given by:

$$i\partial\psi = m\psi \qquad (2.29)$$

we see that Dirac equation is a non-zero-summed theory. From which if Dirac had used this mass conservation equation since mass is equivalent to energy, it will be more difficult for him to justify anti-matter exists within our energy conservation universe. Although experimentally had proven anti-particle as observed within a cloud-chamber [17], my question is that how we can interpret particle traveling backward in time, within a cloud-chamber which is a forward time moving subspace [i.e., temporal ($t > 0$)]. For which we see that, mathematics is not equaled to science since physical significance could have been difference as from purely mathematical standpoint.

2.9 REMARKS

We have seen there is a major difference between science and mathematics; science is physically real, and mathematics is fictitious and virtual. Science is required to exist within our temporal ($t > 0$) universe, but mathematics does not. But without mathematics, science goes nowhere, in which we see that science needs mathematics, but mathematics does not need science. Since science is very complicated, science depends on mathematics to facilitate because mathematics is a language or a video to facilitate the scientific significance and meaning. For this, scientists developed point-singularity approximated equations to describe science, since mathematics is a symbolic representation that simplifies the description of science. Without the point-singularity approximated science, science would be very difficult to learn, to facilitate, and to develop. In this we see that science is a law of approximation, and mathematics is an axiom of absolute certainty. Nevertheless, using exact mathematics to analyze approximated science, it produces inexact solution which may be deviated from the physical reality of science. Since we are scientists but not mathematicians, there is major role difference between them, physical realizable science, and virtual abstract mathematics. In other words, mathematics is a part of that solution, but currently mathematics is also a part of science's problem as virtually fictitious as mathematics is. For this, I have shown that it is not how rigorous and complex mathematics is, it is the physically realizable science that we embrace.

Nevertheless, the essence of this chapter has shown all about physically realizable science. This is dependent on a hypothetical subspace of a postulated model embedded in. For example, if a postulated model is embedded within a timeless (t = 0) subspace, its solution will be as timeless (t = 0) as its embedded subspace. But a timeless solution is not a physically realizable solution that can directly be implemented within our temporal (t > 0) space. Although science is mathematics, it is science that navigates the mathematics, which should not let mathematics lead science.

Since science is also an axiom of logic, without logical thinking, revolutionary science would never have happened. For this, I have seen that we have gradually lost our logical thinking but opted to accept the approval of others. But the burden of science is to be physically realizable; otherwise, it is as virtual as mathematics is since science is mathematics. In other words, any scientific (i.e., mathematical) solution has to comply with the boundary condition of our temporal space, dimensionality, and temporal (t > 0). For instance, Einstein's relativity theories as well as Schrödinger's superposition principle are mathematically correct but physically wrong since they were hijacked by a non-physically realizable empty space paradigm.

REFERENCES

1. F. T. S. Yu, "Time: The Enigma of Space", *Asian Journal of Physics*, 26 (3): 143–158, 2017.
2. F. T. S. Yu, "From Relativity to Discovery of Temporal (t > 0) Universe", *Origin of Temporal (t > 0) Universe: Correcting with Relativity, Entropy, Communication and Quantum Mechanics*, Chapter 1, CRC Press, New York, 1–26, 2019. New York.
3. A. Einstein, *Relativity, the Special and General Theory*, Crown Publisitss, New York, 1961.
4. E. Schrödinger, "An Undulatory Theory of the Mechanics of Atoms and Molecules", *Physical Review*, 28 (6): 1049, 1926.
5. F. T. S. Yu, "What is "Wrong" with Current Theoretical Physicists?", *Advances in Quantum Communication and Information*, Edited by F. Bulnes, V. N. Stavrou, O. Morozov and A. V. Bourdine, Chapter 9, pp. 123–143, IntechOpen, London, 2020.
6. O. Belkind, "Newton's Conceptual Argument for Absolute Space", *International Studies in the Philosophy of Science*, 21 (3): 271–293, 2007.
7. M. Bartrusiok and V. A. Rubakov, *Introduction to the Theory of the Early Universe: Hot Big Bang Theory*, World Scientific Publishing, Princeton, NJ, 2011.
8. R. Zimmerman, *The Universe in a Mirror: The Saga of the Hubble Space Telescope*, Princeton Press, Princeton, NJ, 2016.
9. F. T. S. Yu, "Schrödinger's Cat and His Timeless (t = 0) Quantum World", *Origin of Temporal (t > 0) Universe: Correcting with Relativity, Entropy, Communication and Quantum Mechanics*, Chapter 5, CRC Press, New York, 81–97, 2019. New York.
10. F. T. S. Yu, The Fate of Schrodinger's Cat, *Asian Journal of Physics*, 28 (1): 63–70, 2019.
11. F. T. S. Yu, "Nature of Temporal (t > 0) Quantum Theory: II", *Quantum Mechanics*, Edited by P. Bracken, Chapter 9, pp. 161–188, IntechOpen, London, 2020.
12. C. H. Bennett, "Quantum Information and Computation", *Physics Today*, 48 (10), 24–30, 1995.

13. K. Życzkowski, P. Horodecki, M. Horodecki and R. Horodecki, "Dynamics of quantum entanglement", *Physical Review A*, 65, 1–10, 2001.
14. S. F. Yau and S. Nadids, *The Shape of Inner Space*, Basic Book, NY, 2010.
15. S. Hawking and R. Penrose, *The Nature of Space and Time*, Princeton University Press, New Jersey, 1996.
16. P. A. M. Dirac, "On the Theory of Quantum Mechanics", *Proceedings of the Royal Society A*, 112 (762): 661–677, 1926.
17. N. N. Das Gupta, and S. K. Ghosh "A Report on the Wilson Cloud Chamber and its Applications in Physics", *Reviews of Modern Physics*, 18 (2): 225–365, 1946.

3 Temporal (t > 0) Quantum Theory

Strictly speaking, every scientific solution has to be proven whether it is physical realizable before considering for experimentation, since analytical solution is mathematics. For example, if an elementary particle has proven is not temporal (t > 0), it has no reason to spend that big of a budget for experimentally searching a timeless (t = 0) particle since a timeless particle does not exist within our universe. Similarly, if a mathematician discovers a ten-dimensional subspace, would you not want to prove that their ten-dimensional subspace is a temporal (t > 0) subspace, before experimentally searching for it since a mathematical solution is virtual?

Nevertheless, at the dawn of science, scientists have been using a piece or pieces of papers, drawn models, and paradigms on it and using mathematics as a tool for analyzing a possible solution. But it never occurs to them the background of that piece of paper represented a mathematical subspace that does "not" exist within our universe, for which practically all the laws, principles, and theories were developed from a piece or pieces of papers, which are timeless (t = 0) and, strictly speaking, virtual.

Since science is mathematics but mathematics is "not" equal to science, it is vitally important for us to understand what science really is. In order to understand science, firstly we have to understand what supported the science. For this, the supporter must be the subspace within our universe. In other words, any scientific solution has to be proven to exist within our universe; otherwise it may be fictitious and as virtual as mathematics is, since science is mathematics. In this we see that our universe is a physical subspace that supports every physically realizable aspect within its space, "if and only if" the scientific postulation complies within the existent condition of our universe; dimensionality and causality or temporal (t > 0).

The essence of our temporal (t > 0) universe is that if a mathematical solution is "not" complied with within the temporal (t > 0) condition of our universe, it cannot exist within our universe. Since quantum mechanics is one of the pillars of modern science, I will start with one of the most intriguing principles in quantum mechanics, the uncertainty principle. I will carry the principle onto a newly found "certainty" principle. In this I will show Heisenberg's principle was based on a diffraction limited observation, instead of upon a "nature" of time, developing his principle. I will also show the mystery of coherence theory can be understood with a principle of certainty. In this I will show that certainty subspace can be created within our temporal (t > 0) universe. Samples applied to synthetic aperture imaging and wave front reconstruction will be included.

DOI: 10.1201/9781003271505-3

3.1   SCIENCE AND MATHEMATICS

There is a profound relationship between science and mathematics, in which we have seen that without mathematics there would be no science. In other words, science needs mathematics but mathematics does not need science. Although science is mathematics, mathematics is not science. For example, if any mathematical solution cannot be proven it exists within our universe, then its solution is "not" a "physical realizable" solution that can be "directly" implemented within our temporal (t > 0) universe.

But this is by no means to say that solutions are not temporal (t > 0) or timeless (t = 0) solutions are not science. In fact, practically all the fundamental laws, principles, and theories are timeless (t = 0) or time independent. And these timeless (t = 0) laws, principles, and theories were and "still" are the cornerstone and foundation of our science, as I will call them timeless (t = 0) or time-independent science, a topic I will elaborate on in a different occasion. For simplicity, let me take one of the simplest examples, Einstein's energy equation [1], as given by:

$$E = mc^2 \qquad\qquad (3.1)$$

where E is the energy, m is the mass, and c is the velocity of light. This equation is one of the most famous equations in science, yet it is timeless (t = 0). Although this equation has been repeatedly used and applied in practice, but strictly speaking it cannot be directly implemented within our temporal (t > 0) universe, since it is not a time-variable function. Let us transform Einstein's equation into a time-variable equation as given by [2]:

$$\frac{\partial E(t)}{\partial t} = -c^2 \frac{\partial m(t)}{\partial t}, \quad t > 0 \qquad\qquad (3.2)$$

where $\partial E(t)/\partial t$ is the rate of increasing energy conversion, $-\partial m/\partial t$ is the corresponding rate of mass reduction, c is the speed of light, and t > 0 denotes a forward time-variable equation. In this we see Eq. (3.2) is a time-dependent equation that exists at time t > 0, which represents a forwarded time variable function that only occurs after time excitation at t = 0. Incidentally, this is the well-known "causality" constraint (i.e., t > 0) [3] as imposed by our temporal (t > 0) universe.

Nevertheless, every mathematical postulation needs to prove that there exists a solution before we search for it, although it is not guarantee that we can find it. But it seems to me it does not have a criterion to prove that a hypothetical scientific theory or principle existed within our universe before we actually experimentally confirm it. For example, an analytical solution tells us if there is an "angle particle" derived from a complicated mathematical analysis, would you not want to find out the solution existed within our temporal (t > 0) universe before experimentally searching for it? And this is precisely what we shall know first before experimentation has taken place, since it is very costly in time and in revenue to find a physical particle.

Although science needs mathematics, without simple mathematical approximation, science would be very difficult to learn and to facilitate. And this is precisely the reason practically all the fundamental laws are point-singularity approximated from which we see that all the laws, principles, and theories are approximated. Again, we take Einstein's energy equation of Eq. (3.1) as an example; we see that this equation has no dimension and size, and it is a typical point-singularity approximated equation. It is discernible if we include all the negligibly terms; "physical significances" of this equation would be overwhelmed by the complexity of mathematics. For this we see that an ounce of good approximation is worth more than tons of mathematical illustrations!

Let me stress that the essence of simplicity in science is that without the symbolic substitution and approximation, it will be extremely difficult or even impossible to develop science since science itself is already very complicated. Yet the simplicity representation of science also has its own drawback, referred as "classical and deterministic (i.e., classical science). The implication of deterministic or classical is totally misled by our part since our predecessors who developed those laws and principles "precisely" understood the deficiency of approximation. Yet without the approximated presentation, how can we develop science? Instead of ignoring our predecessors' wisdom, turns around we had treated them as "deterministic" or classical, which were "never" been our predecessor's intention. Again, without the point-singularity approximated science, please tell me how we can develop those simple and elegant laws, principles, and theories. Although those laws, principles, and theories are timeless $(t = 0)$, most of them were and "still" are the cornerstones of our science. Nevertheless, mathematics is a "symbolic" langue of science, but mathematics is not science.

Since all laws, principles, and theories were made to be broken or revised or even to beeradicated, as science advances into sub-subatomic scale regime and moving closer to real-time transmission, those timeless $(t = 0)$ laws, principles, and theories produced incomprehensible consequences, particularly as applied directly within our temporal $(t > 0)$ universe. For example, such as applied to superposition principle to quantum computing and communication, superposition is a timeless $(t = 0)$ principle [4].

3.2 TEMPORAL (T > 0) SUBSPACE

In this section, I will show several subspaces that have been used in the past, as depicted in Figure 3.1, since any physically realizable law, principle, and theory depends on their physically realizable paradigm. In this we see that it is the paradigm that dictates the physically realizable science but "not" the severity of rigorous mathematics.

For example, when you are designing a submarine, the subspace that the submarine is supposed to be situated within is vitally important; otherwise, your submarine will very "likely" not survive thousands of feet of underwater pressure. Therefore, it is necessary to know the subspace that a postulated science is implementing into it; otherwise, the postulated science very likely "cannot" exist within the subspace.

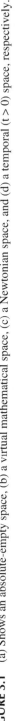

(d)

t > 0
f(x, y, z; t)

Yu's temporal subspace;
has coordinates,
non-empty,
Time-dependent space

(c)

-∞ < t < ∞
f(x, y, z)

Newtonian subspace;
has coordinates,
non-empty,
Time-independent space.

(b)

t = 0
f(x, y, z)

Virtual subspace;
has coordinates, empty,
Timeless space.

(a)

t = 0

Absolute-empty Timeless
subspace;
has no coordinate.

FIGURE 3.1 (a) Shows an absolute-empty space, (b) a virtual mathematical space, (c) a Newtonian space, and (d) a temporal (t > 0) space, respectively.

From Figure 3.1 we see that there is an absolute-empty space, a mathematical virtual space, a Newtonian's space, and a temporal (t > 0) space. An absolute-empty space or just empty space has no substance and has no time. A mathematical virtual space is an empty space that has no substance in it, but mathematicians and theoretical scientists can implant a coordinate system in it, since mathematics is virtual and theoretical scientists are also mathematicians.

We note that mathematical virtual space has been used over centuries by scientists at the dawn of science, but this is a virtual space that does "not" exist within our temporal (i.e., t > 0) universe. The next subspace is known as Newtonian space [5]; it has substance and coordinates in it, but treated time as an "independent" variable, for which Newtonian and mathematical spaces are virtually the "same". Since Newtonian space is time independent, it "cannot" exist within our temporal (t > 0) space since time and substance have to "mutually coexist" within our temporal (t > 0) universe. Yet scientists have been using Newtonian space for their analyses over centuries and not knowing it is a virtual space.

The last subspace is known as temporal (t > 0) space [6], where time and substance interdependently "coexist" and time is a forward "dependent variable" that runs at a "constant speed". We stress that this temporal (t > 0) subspace is currently "only" a physically realizable subspace. Notice that Einstein's energy equation [2] was developed from his 4-dimensional spacetime continuum, but our temporal (t > 0) space is a time dependent space (see Appendix E), where time is not an independent variable.

Physical reality is that any scientific hypothesis that deviates "away" from the boundary condition that imposed by our temporal (t > 0) universe is "not" a physically realizable solution. But this is does not mean that the virtual mathematical empty space and Newtonian space are useless. The fact is that all the laws, principles, and theories were developed within timeless (t = 0) or Newtonian subspaces "inadvertently", at the dawn of science. In fact it was from the background subspace of a piece of paper although not intentionally [7], for which practically all the laws, principles, and theories are timeless (t = 0).

Nevertheless, what temporal (t > 0) space means is that any subspace coexists with time, where time is a forward-dependent variable with respect to its subspace and its speed had been well settled when our universe was created. This means that before the creation of our temporal universe, there is a "larger" temporal space that our universe is embedded in; otherwise, our universe will "not" be exited. Nevertheless, every subspace within our universe is a time-varying stochastic [8] subspace, in which every substance or subspace changes with time but does not change time. Strictly speaking, our universe is a "temporal (t > 0) stochastic expanding subspace"; any postulated law, principle, and theory has to comply with the temporal (t > 0) condition within our universe; otherwise, it is as virtual as mathematics.

3.3   TIMELESS (T = 0) SPACE

Let me show what mathematicians can do within a virtual subspace, as depicted in Figure 3.2. Since quantum mechanists are also mathematicians, they can implant a coordinate system within an empty space as they wish, regardless of whether the model is physically realizable or not.

(a)   (b)

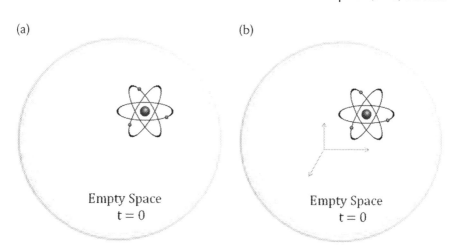

FIGURE 3.2 Shows a set of atomic models embedded within virtual empty subspaces. (a) Shows a singularity approximated atomic model is situated within an empty space, which has no coordinate system. (b) Shows an atomic model is embedded withinan empty space that has a coordinate system drawn into it.

The basic difference between Figure 3.2(a) and Figure 3.2(b) is that there is a virtual coordinate system has been added in Figure 3.2(b) by quantum mechanists. Once the coordinate system is implanted, dimensionality of the sub-atomic particles cannot be ignored. The reason is that for the atomic model to exist within the subspace, the atomic model has to "comply" with the existence conditions within the subspace, since it is the subspace that affects the solution and not the solution that changes the subspace. In this we see that neither Figure 3.2(a) nor Figure 3.2(b) are "not" physically realizable paradigms. For this solution obtained from these empty subspace models, it will be timeless (t = 0) or time independent, since timeless and time independent are equivalent.

Aside from the non-physically realizable paradigms of Figure 3.2(a), I will show what a timeless (t = 0) subspace can do for substances within the subspace. Let me assume we have three particles that are situated within an empty space, as normally done on a "piece of paper", shown in Figure 3.3.

Since an empty subspace has "no time", all particles within the subspace collapse or "superimpose" instantly all together at t = 0, because time is distance and distance is time. This is precisely what the "simultaneous and instantaneous" superposition principle does in quantum mechanics [4]. The reason particles collapsed at t = 0 is because the subspace has "no time". And the other reason that particles are superimposed together, is because within a timeless (t = 0) space, they have "no distance "or space.

By virtue of energy conservation, we see that superimposed particles have a mass equal to the sum of entire superimposed particles, but they have "no size". In view of timelessness space, we see that superimposed particles can be found everywhere within the entire timeless (t = 0) subspace, since timeless (t = 0) subspace has "no" distance, as depicted hypothetically in Figure 3.4, from which

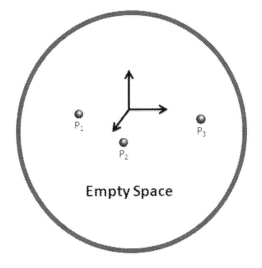

FIGURE 3.3 A hypothetical scenario shows three particles are embedded within an empty subspace.

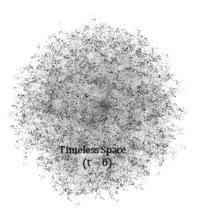

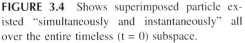

FIGURE 3.4 Shows superimposed particle existed "simultaneously and instantaneously" all over the entire timeless (t = 0) subspace.

we see that Schrödinger's fundamental principle of superposition existed within a virtual timeless (t = 0) subspace, and it cannot exist within our temporal (t > 0) universe, since timeless and temporal are "mutually exclusive".

By the way, this is precisely the superposition principle that Einstein was objecting to, which he called spooky. As I quote from a 1935 *New York Times*' article (i.e., Figure 3.5): "Einstein and two scientists found quantum theory is incomplete even though correct" [9]. In view of the preceding illustration, we see that Schrödinger's superposition principle is "correct" but only within a timeless (t = 0) subspace and it is "incorrect" within our temporal (t > 0) space" since timeless space cannot exist within our temporal universe (i.e., TEP).

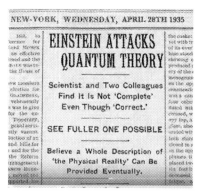

FIGURE 3.5   A 1935 *New York Times'* article.

3.4   TIME IS NOT AN ILLUSION BUT REAL

As we accepted subspace and time coexist within our temporal (t > 0) universe, time has to be real and it cannot be virtual, since we are physically real. And every physical existence within our universe is real. The reason some scientists believed time is virtual or an illusion is that it has no mass, no weight, no coordinate, no origin, and it cannot be detected or even be seen. Yet time is an everlasting, existing, real variable within our known universe. Without time there would be no physical matter, no physical space, and no life. The fact is that every physical matter is associated with time, including our universe. Therefore, when one is dealing with science, time is one of the most enigmatic variables that is ever present and cannot be simply ignored. Strictly speaking, all the laws of science as well as every physical substance cannot exist without the existence of time. For this we see that time "cannot" be a dimension or an illusion. In other words, if time is an illusion, then time will be "independent" from physical reality or from our universe. And this is precisely what many scientists have treated time as an "independent" variable, such as Murkowski's space [1], for which the space can be "curved" or time-space can be changed by gravity [10]. If space can be curved, then we can change the "speed" of time. In other words, does our universe exist with time, or does time exist with the universe? The answer is our universe exists with time, although space and time coexist.

As time coexists with subspace, we see that any subspace within our temporal (t > 0) universe cannot be empty, and the speed of time is the same everywhere within our universe. This means that the speed of time within a subspace is "relative" with respect to the different subspaces, as based on Einstein's special theory of relativity [1]. For example, subspaces closer to the edge of our universe run faster "relative" to ours, but the speed of time within the subspaces near the edge as well within our subspace are the "same", which has been determined by the speed of light as our universe was created by a big bang theory using Einstein's energy equation as given by [6]:

$$\frac{\partial E(t)}{\partial t} = -c^2 \frac{\partial m(t)}{\partial t}, \quad t > 0 \tag{3.3}$$

where $\partial E/\partial t$ is the rate of increasing energy conversion, $-\partial m/\partial t$ is the corresponding rate of mass reduction, c is the speed of light, and $t > 0$ represents a forward-time variable or temporal. In this we see that a "time-dependent" equation exists at time $t > 0$; a well-known causality constraint (i.e., $t > 0$) [3] is imposed by our universe. Similarly, the preceding equation can be written as:

$$\frac{\partial E}{\partial t} = -c^2\frac{\partial m}{\partial t} = [\nabla\cdot S(v)] = -\frac{\partial}{\partial t}\left[\frac{1}{2}\varepsilon E^2(v) + \frac{1}{2}\mu H^2(v)\right], \quad t > 0 \qquad (3.4)$$

where ε and μ are the permittivity and the permeability of the deep space, respectively; v is the radian frequency variable; $E^2(v)$ and $H^2(v)$ are the respective electric and magnetic field intensities; the negative sign represents the "outflow" energy per unit time from an unit volume; $(\nabla\cdot)$ is the divergent operator; and S is known as the Poynting vector or "energy vector" of an electromagnetic radiator, as can be shown by $S(v) = E(v) \times H(v)$ [11].

In view of this equation, we see how our universe was created, as depicted by a composite diagram in Figure 3.6, in which we see that radian energy (i.e., radiation) diverges from the mass, as mass reduces with time. In this we see that our universe enlarges and its boundary expands at the speed of light.

Figure 3.7 shows a schematic diagram of our temporal ($t > 0$) universe, which depicts approximately the behavior of subspace changes as its boundary expands

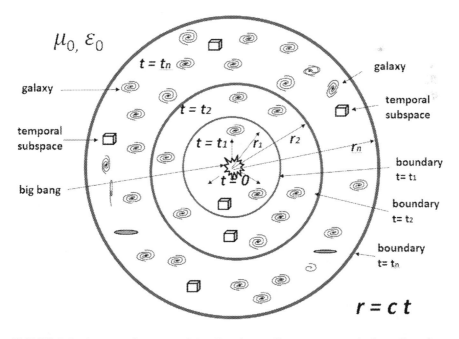

FIGURE 3.6   A composite temporal ($t > 0$) universe diagram. $r = ct$, r is the radius of our universe, t is time, c is the velocity of light, and ε_0 and μ_0 are the permittivity and permeability of the space.

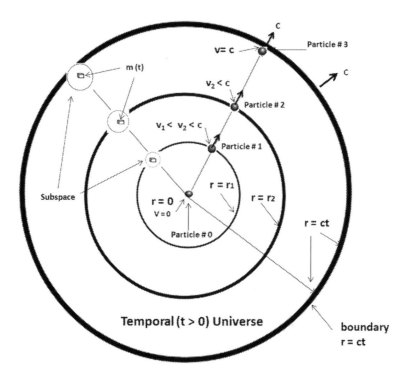

FIGURE 3.7   Shows a schematic diagram of our temporal (t > 0) universe. c is the speed of light, m(t) is the temporal mass, v is the radial velocity.

with the speed of light. In this we see that the subspace enlarges faster closer toward the boundary, but a solid substance m(t) changes little within the subspace. We also see that the outward speed of the particle (or subspace) increases "linearly" as the boundary increases with light speed. For example, outward speed of particle 2 is somewhat faster than particle 1 (i.e., $v_2 > v_1$). For this we see that our universe is a dynamic temporal (t > 0) "stochastic" universe; a simple geometrical equation or mathematical abstract space cannot describe this. One of the important aspects of our universe is that every subspace, no matter how small it is, "cannot" be empty and it has time.

For instance, in order for us to exist within our planet, we must be temporal (t > 0); that is, we have time, and we must change naturally with time, otherwise we cannot exist within our universe. In other words, our time is the same as our planet and the universe, but the velocity of our planet is different from other subspaces. For example, subspaces near the edge of our universe are moving faster than us, for which it has "relative" speed of time between us and a subspace closer to the edge of our universe as based on special relativity. Yet within our universe the pace of time is the same everywhere. On the other hand, if we assumed that we are timeless (t = 0), we could "not" have existed within our universe, since time and timelessness are mutually exclusive!

I further note that any subspace within our universe cannot be empty, since a subspace coexists with time. Although a subspace coexists with time, time is not equal to a subspace. Yet, space is time and time is space since time and space are mutually inclusive. For example, a substance has dimension (or space), but time has no dimension and no mass. In this we see that time is "not" an independent dimension, but it "dependently" exists with respect to a subspace. For this we stress that it is our universe that governs the science and it is not the science that changes our universe.

Once again, we have shown that time is "not "an illusion or virtual, time is physically real; because everything that exists within our living space is physically real, otherwise it will not exist within our temporal universe. In other words, everything within our universe is temporal (t > 0), of which I have discovered that practically all the laws, principles, theories, and paradoxes of science were developed from a timeless (t = 0) platform (i.e., a piece or pieces of papers) for centuries, at the dawn of science "inadvertently" [7].

Nevertheless, one important aspect within our universe is that every subspace has a price; an amount of energy ΔE and a section of time Δt to create (i.e., ΔE and Δt), and it is "not free". For example, a simple facial tissue takes a huge amount of energy ΔE and a section of time Δt to create. It is however a "necessary" but not a sufficient condition, because it also needs an amount of information ΔI to make it happen (i.e., ΔE, Δt and ΔI) [12].

In short, I would stress that if there is a beginning then there is an end. Since time and space coexist, then time and space have no beginning and no end. In this we see that time-space [or temporal (t > 0) space] is ever existed, since existence and non-existence are mutually exclusive. In other words, if emptiness and non-emptiness are mutually excluded, then time always exists with space. Thus, time is real because the space is real, for which time-space has no beginning and has no end. And this must be the art of the temporal (t > 0) universe.

3.5   LAW OF UNCERTAINTY

One of the most intriguing principles in quantum theory [13] must be Heisenberg's uncertainty principle [14], as shown by the following equation:

$$\Delta p \ \Delta x \geq h \tag{3.5}$$

where Δp and Δx are the momentum and position errors, respectively; and h is Planck's constant. As reference to "wave-particle dynamics", the momentum p of a "photonic particle" is presented by a "quanta" of energy $h\upsilon$, as given by:

$$p = h/\lambda = h\upsilon/c \tag{3.6}$$

where h is Planck's constant, λ is the wavelength, υ is the frequency, and c is the velocity of light. In this we see that Heisenberg's principle was based on "wave-particle duality" that existed within an "empty space". The essence of Heisenberg's uncertainty principle is that one cannot precisely determine the position x and the momentum p of a particle "simultaneously under observation", as illustrated in Figure 3.8. In this the principle is time "independent", since Heisenberg's principle

 v

FIGURE 3.8 A particle in motion within an "empty" subspace. v is the velocity.
Note: Background paper has been treated as an "empty" subspace for centuries.

A piece of paper

was based on an "observation" standpoint that has nothing to do with changing naturally with time. Yet we know that if there is "no" time there is "no" uncertainty.

With reference to Figure 3.8, we see that Heisenberg's principle was basically derived on an empty timeless (t = 0) subspace, and it has nothing to do or independent with the underneath subspace that the particle is situated. Strictly speaking, it is not a physically realizable paradigm that should be used in the first place, since a particle and empty subspace are mutually exclusive. Secondly, the position error Δx of Heisenberg was based on a diffraction-limited microscopic observation, where the spatial ambiguity of Δx is given by [15]:

$$\Delta x = 0.6 \; \lambda / \sin \alpha \qquad (3.7)$$

where λ is the observation wavelength, $2(\sin \alpha)$ is the "numerical aperture" of the microscope, and α is the subtended half-angle of observation aperture. In this we see that the position error Δx is "not" due to a particle in motion but is based on the diffraction-limited aperture. This is precisely why Heisenberg's position error Δx has been interpreted as an "observation error", which is independent with time. But uncertainty changes naturally with time, since without time it has no uncertainty.

Secondly, the momentum error Δp is as I quoted [16]: After collision, the particle being observed; the photon's path is only to lie within a cone having semi-vertical angle α in which the momentum of the particle is uncertain by the amount as given by:

$$\Delta p = h(\sin \; \alpha)/\lambda \qquad (3.8)$$

where λ is the wavelength of the quantum leap of hυ. From this we see that momentum error Δp is not due to bandwidth $\Delta \upsilon$ of quantum leap since any physical radiator has to be band limited. In other words, the momentum error Δp of the preceding Eq. (3.4) is a singularity approximated λ, which is not a band-limited $\Delta \lambda$ of physical reality.

As we look back at the subspace that Heisenberg's principle developed from, it was an inadvertently timeless (t = 0) subspace, as shown in Figure 3.8. Aside from the timeless (t = 0) subspace, it is the uncertainty mainly due to diffraction-limited observation that is a secondary cause by human intervention, but not due to a natural change with time. This is similar to the entropy theory of Boltzmann [17]: Entropy increases naturally with time within an enclosed subspace. For this uncertainty should be increasing naturally with time, without human intervention. As I have noted, without time, there would be no entropy and no uncertainty.

Nevertheless, momentum error Δp and position error Δx are mutually "coexisting". In principle, they can be traded. But the trading cannot be without constraint since time is a dependent-forward variable. But Heisenberg's uncertainty Δp and Δx are mutually independent, since this position error Δx is due to diffraction-limited observation, which has nothing to do with time. For this it poses a physical inconsistency within our temporal (t > 0) universe, although Heisenberg's principle has been widely used without any abnormality. But it is from the physical consistency standpoint that Heisenberg's position error Δx was based on a diffraction-limited observation that has nothing to do with time. And also this momentum error Δp was based on a singularity wavelength λ, which is not a band-limited reality.

Yet, the uncertainty principle can be made temporal (t > 0), similar to the entropy theory of Boltzmann. For this we have a law of uncertainty as stated: Uncertainty of an isolated particle increases naturally with time. Or more specifically: Uncertainty of an isolated particle within an isolated subspace increases with time and eventually reaches a maximum amount within the isolated subspace. For this we see that there exists a profound connection between uncertainty and entropy.

3.6   TEMPORAL (T > 0) UNCERTAINTY

It is our universe that governs the science, and it is not the science that governs our universe. Therefore, every principle within our universe has to comply with the temporal (t > 0) condition within our universe; otherwise, the principle cannot exist within our universe. This includes all the laws, principles, and theories, such as Maxwell's electromagnetic theory, Boltzmann's entropy theory, Einstein's relativity theory, Bohr's atomic model, Schrödinger's superposition principle, and others. This uncertainty principle cannot be the exception.

Let us now assume a temporal (t > 0) particle m(t) is situated within a temporal (t > 0) subspace, as depicted in Figure 3.9. Strictly speaking, any particle that exists

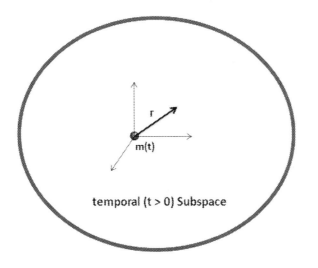

FIGURE 3.9   Shows a temporal (t > 0) particle m(t) within a temporal (t > 0) subspace. r is the radial direction.
Note: It is a physically realizable paradigm since a temporal particle m(t) is embedded within a temporal subspace.

within a temporal subspace must be a temporal (t > 0) particle; otherwise, the particle cannot exist within our temporal (t > 0) universe.

For simplicity, we further assume m(t) has no time or is pseudo-timeless; after all, science is a law of approximation. The same is for Heisenberg's assumption: the particle is a photonic particle (i.e., a photon); as from the wave particle-duality standpoint [18] (see Appendix B), momentum of a photon is given by:

$$p = h/\lambda = h\ \upsilon/c \qquad\qquad (3.9)$$

where h is Planck's constant, λ is the wavelength, and υ is the frequency of the photonic particle. As I have mentioned earlier, within our universe any radiator has to be band limited. Thus, the momentum error is naturally due to changes in bandwidth $\Delta\upsilon$, as given by:

$$\Delta p = h\ \Delta\upsilon/c \qquad\qquad (3.10)$$

Instead of using a cone of light, as Heisenberg had postulated. By virtue of the time-bandwidth product $\Delta\upsilon\ \Delta t = 1$, $\Delta\upsilon$ decreases with time. This position error can be written as:

$$\Delta r = c\ \Delta t \qquad\qquad (3.11)$$

where r is the radial distance, we have the following uncertainty relationship.

$$\Delta p\ \Delta r = [h\ \Delta\upsilon/c][c\ \Delta t] = h\ \Delta\upsilon\ \Delta t \qquad\qquad (3.12)$$

In this we see that $\Delta\upsilon \cdot \Delta t$ is the time-bandwidth product. As we imposed the optimum energy transfer criterion on the time-bandwidth product [12], we have the following relationship:

$$\Delta\upsilon\ \Delta t \geq 1 \qquad\qquad (3.13)$$

Since the lower bound for a photonic particle is limited by Planck's constant, we have the following equivalent form as given by:

$$\Delta E\ \Delta t \geq h \qquad\qquad (3.14)$$

Nevertheless, in view of Eq. (3.13), the momentum uncertainty principle can be shown as:

$$\Delta p\ \Delta r \geq h, \quad t > 0 \qquad\qquad (3.15)$$

where (t > 0) denotes that the uncertainty principle is complied with the temporal (t > 0) condition within our universe. In view of either the conservation of momentum or equivalently energy conservation we see that position error Δr increases naturally with time, which shows that momentum error Δp decreases naturally with bandwidth $\Delta \upsilon$, as in contrast with Heisenberg's assumption, momentum error Δp has "nothing" to do with the changes of $\Delta \upsilon$. This is precisely the law of uncertainty that I have described earlier; uncertainty of an isolated particle increases naturally with time.

Since the increases in position error Δr are due to time, it must be due to the dynamic expansion of our universe [6]. For example, as the boundary of our universe is constantly expanding at the speed of light, by virtue of energy conservation, every aspect within our universe changes naturally with time as our universe changes. As time moves on naturally, the larger the position error Δr increases with respect to the starting point, as illustrated in Figure 3.10.

Therefore we see that uncertainty is not a static process but a temporal (t > 0) dynamic principle, as in contrast with Heisenberg's position error Δr that is independent with time and the momentum error Δp is independent from $\Delta \upsilon$. In this we see that if there is no time (i.e., an empty space), there is no uncertainty and no probability. Nevertheless, each of the uncertainty units or cells, $(\Delta p, \Delta x)$, $(\Delta E, \Delta t)$, and $(\Delta \upsilon, \Delta t)$ are self-contained. In other words, ΔE and Δt coexist and they can be bilateral traded, but under the constraint of time as a forward-moving dependent variable. For example, if a section of Δt has been used, we cannot get the identical section of time back, but we can exchange for a different section of Δt. In this we see that we can trade for a narrower Δt with a wider ΔE or wider Δt with a narrower ΔE, but we cannot trade (i.e., Δt for ΔE), since Δt represents a section of time that has no physical substance to manipulate. In other words, it is that momentary section of time that has been used (i.e., expended) that we cannot get it back, since time is a forward variable.

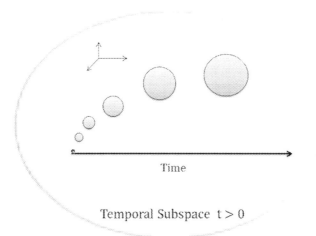

FIGURE 3.10 Shows position error Δr (i.e., sphere of Δr) enlarges naturally with time within a temporal (t > 0) subspace: Δr represents a position error of the particle; at various locations as time moves constantly.

3.7   CERTAINTY PRINCIPLE

One of the important aspects of temporal uncertainty is that the subspace within our universe is a temporally (t > 0) uncertain subspace. In other words, every subspace is a temporal (t > 0) stochastic subspace, such that the dynamic behavior of the subspace changes dependently with time. For this, any changes within our universe have a profound connection with the constantly expanding universe. In this we have shown that uncertainty increases naturally with time, even without any other perturbation or human intervention. Similar to the myth of Boltzmann's entropy theory [17], entropy increases naturally with time within an enclosed subspace, which has been shown to be related to the expanding universe [6].

Similarly, there is a profound connection between coherence theory [15] and certainty principle as I shall address. Nevertheless, it is always a myth of coherence, as referred to in Figure 3.11, where the coherence theory can be easily understood by Young's experiment. In this the degree of coherence can be determined by the visibility equation as given by:

$$\nu = \frac{\text{Imax} - \text{Imin}}{\text{Imax} + \text{Imin}} \tag{3.16}$$

where I_{max} and I_{min} are the maximum and minimum intensities of the fringes. But the theory does not tell us where the physics come from. For this it can be understood from the certainty principle as I shall address.

It is trivial that if there is an uncertainty principle, it is inevitable to not have a certainty principle. This means that as the photonic particle we are looking for is likely to be found within a certainty subspace. Perfect certainty (or absolute uncertainty) occurs at $t = 0$ (i.e., present instant moment), which is a timeless ($t = 0$) virtual subspace that does not exist within our temporal (t > 0) universe. Nevertheless, the certainty principle can be written in the following equivalent forms:

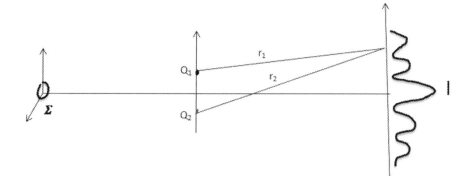

FIGURE 3.11   Young's experiment. Σ represents an extended monochromatic source, Q_1 and Q_2 are the pinholes, and I represents the irradiance distribution.

$$\Delta p \ \Delta r < h, \quad (t > 0) \tag{3.17}$$

$$\Delta E \ \Delta t < h, \quad (t > 0) \tag{3.18}$$

$$\Delta v \ \Delta t < 1, \quad (t > 0) \tag{3.19}$$

where $(t > 0)$ denotes that the equation is subjected to temporal $(t > 0)$ constraints. In view of the position error Δr in Eq. (3.17), it means that it is likely the photonic particle can be found within the certainty subspace. Since the size of the subspace is limited by Planck's constant, h, it is normally used as a limited boundary not to be violated. Yet within this limited boundary, the certainty subspace had been exploited by Dennis Gabor for his discovery of wave front reconstruction in 1948 [19] and it was applied to synthetic aperture radar imaging in 1950s [20]. Yet there is a price to pay, namely a section of time Δt, as we will see in a moment.

Since the size of certainty subspace is exponentially enlarging as the position error Δr increases, for which the "radius" of the certainty sub-sphere is given by:

$$\Delta r = c \ \Delta t = c/(\Delta v) \tag{3.20}$$

where c is the speed of light, Δt is the time error, Δv is the bandwidth of a light source or a quantum leap hv, and v is the frequency. Thus, we see that position error Δr is inversely proportional to bandwidth Δv, as plotted in Figure 3.12.

In view of this plot, we see that when bandwidth Δv decreases, a larger certainty subspace enlarges exponentially since the volume of the subspace is given by:

$$\text{Certainty subspace} = (4\pi/3)(\Delta r)^3 \tag{3.21}$$

In this we see that a very "large" certainty subspace can be realized within our universe, which is within a limited Planck's constant, h, as depicted in Figure 3.13, where we see a steady-state radiator A emits a continuous band-limited Δv electromagnetic wave, as illustrated. A certainty subspace with respect to an assumed

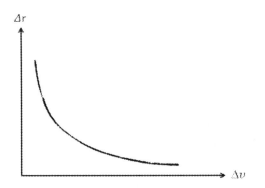

FIGURE 3.12 A plot of position error Δr versus bandwidth Δv.

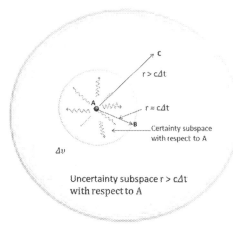

FIGURE 3.13 Shows a certainty subspace is embedded within uncertainty subspace. A is assumed as a steady-state photonic particle emits a band-limited Δv radiation; r is the radius with respect to the emitter A; and B represents the boundary of certainty subspace of A.

photonic particle A for a given Δt can be defined as illustrated within $r = c \, \Delta t$, where $\Delta t = 1/\Delta v$. In other words, it has a high degree of certainty to find particle A within the certainty subspace. Nevertheless, from a electromagnetic disturbance standpoint, the certainty subspace provides a high degree of certainty (i.e., degree of coherence) with respect to point A.

As from the coherence theory standpoint, any other disturbances away from point A but within the certainty subspace (i.e., within $r < c \, \Delta t$) are mutual coherence (i. e., certainty) with respect to A, where $r = c \, \Delta t$ is the radius of the certainty subspace of A. In other words, any point-pair within $d < c \, \Delta t$, where $\Delta t = 1/\Delta v$, is mutual coherence within a radiation subspace. On the other hand, a distance greater than $r > c \, \Delta t$ from point A is a mutually uncertain subspace with respect to A. In other words, any point-pair distance larger than $d > c \, \Delta t$ within the radiation space is mutually incoherent. In this we see that it is more unlikely to find the same photonic particle after it has been seen at point A outside A's certainty subspace.

Since the certainty subspace represents a "global" probabilistic distribution of a particle's location from a particle physicist's standpoint, this means that it is very likely the particle can be found within the certainty subspace. In this we see that a postulated particle firstly is temporal $(t > 0)$ or has time; otherwise there is no reason to search for it. Then after it has been proven it is a temporal [i.e., $m(t)$] particle, it is more favorable to search the particle within a certainty subspace.

The essence of wave-particle duality is a mathematically simplistic assumption to the equivalence of a package of wavelet energy as equivalently to a particle in motion from a statistical mechanics standpoint, in which the momentum $p = h/\lambda$ is conserved. However, one should not treat a wave as equal to a particle or a particle as equal to a wave. It is the package of wavelet energy equivalent to a particle dynamic (i.e., photon), but they are not equal. Similar to Einstein's energy equation, mass is equivalent to energy and energy is equivalent to mass, but they are not equal. Therefore, as from energy conservation we see that bandwidth $\Delta \upsilon$ decreases with time is the physical reality instead of treating a package of wavelets as a particle (i.e., photon), that was due to the classical mechanics standpoint that has treated the quantum leap momentum $p = h/\lambda$. In this we see that photon is a virtual particle, although many quantum scientists have regarded a photon as a physical particle.

We further note that any point-pair within the certainty subspace exhibits some degree of certainty or coherence, which has been known as mutual coherence [15]. And the mutual coherence can be easily understood, as depicted in Figure 3.14, in which we see that a steady-state band-limited $\Delta \upsilon$ electromagnetic wave is assumed to exist within a temporal (t > 0) subspace. As we pick an arbitrary disturbance at point B, a certainty subspace of B can be determined within $r \leq c\,\Delta t$, as shown in the figure. This means that any point disturbance within the certainty subspace has a strong certainty (or coherence) with respect to a point B disturbance. Similarly, if we pick an arbitrary point A, then a certainty subspace of A can be defined as illustrated in the figure, of which we see that a portion is overlapped with the certainty subspace of B. Any other disturbances outside the corresponding subspaces of certainty A, B, and C are the uncertainty subspaces. It is trivial to see that a number of configurations of certainty subspaces can be designed for application. In this we see that a multi-wavelength configuration, such as $\Delta \upsilon_1$, $\Delta \upsilon_2$ and $\Delta \upsilon_3$, can also be simultaneously implemented to create various certainty subspaces, such as for multi-spectral imaging or information processing application.

One commonly used for producing a certainty subspace for a complex wave front reconstruction is depicted in Figure 3.15 [21]. In this we see that a band-limited $\Delta \upsilon$ laser is employed, where a beam of light is split up by a splitter BS. One beam, B_2, is directly impinging on a photographic plate at plane P and the other

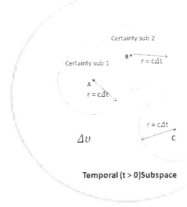

FIGURE 3.14 Shows various certainty subspace configurations, as with respect to various disturbances within a steady-state band-limited $\Delta \upsilon$ electromagnetic environment within a temporal (t > 0) subspace.

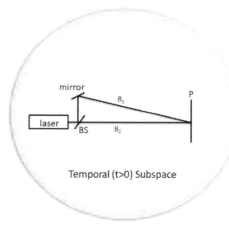

FIGURE 3.15   Shows an example of an exploiting certainty subspace for wave front re-construction. BS: beam splitter, P: photographic plate.

beam B_1 is diverted by a mirror and then is combined with beam B_2 at the same spot on the photographic plate P. It is trivial to know that if the difference in distances between these two beams is within the certainty subspace, then B_1 and B_2 are mutually coherent (or certain); otherwise they are mutually incoherent (or uncertain). In this we see that the distance between B_1 and B_2 is required as given by:

$$|d_1 - d_2| < c \ \Delta t = c/\Delta v \tag{3.22}$$

where d_1 and d_2 are the distances of bean B_1 and B_2, respectively, from the splitter BS. In this we see that the radius of the certainty subspace of BS is written by:

$$\Delta r = |d_1 - d_2| < c \ \Delta t = c/\Delta v \tag{3.23}$$

where $|d_1 - d_2| = c/\Delta v$ is the coherent length of the laser. In this we see that by simply reducing the bandwidth Δv, a larger certainty subspace can be created within a temporal (t > 0) subspace.

3.8   ESSENCE OF CERTAINTY PRINCIPLE

Since every substance or subspace within our universe was created by an amount of energy ΔE and a section of time Δt [i.e., $(\Delta E, \Delta t)$], any changes of ΔE changes the size of certainty subspace Δr. This is a topic that astrophysicists may be interested. Similar to particle physicists, subatomic particle have to be temporal (t > 0); otherwise the particle must be a virtual particle that cannot exist within our universe. Secondly, it is more likely a temporal (t > 0) particle is found within its certainty subspace; otherwise it will be searching a timeless (t = 0) particle "forever" within our temporal (t > 0) universe. In view of the certainty unit, ΔE and Δt mutually coexist in which time is a forward-dependent variable. Any changes of ΔE

can only happen with an expenditure of a section of time Δt, but it cannot change the speed of time. Since the energy is conserved, Δt is a section of time required to have the amount of ΔE within a certainty unit of $(\Delta E, \Delta t)$. In other words, ΔE and Δt can be traded; for example, a wider variance of ΔE is traded for a narrower Δt.

Nevertheless, time has been treated as an independent variable for decades, as normally assumed by scientists. But whenever a section of time Δt has been used, it is not possible to bring back the original moment of Δt, even though it is possible to reproduce the same section of Δt. This is similar if we reconstruct a damaged car, but we cannot bring back the original car that has crashed. And this is precisely the price of time to pay for everything within our universe. Then my question is that if time is a forward-dependent variable with respect its subspace, how can we "curve" the space with time? Similarly, we coexist with time, so how can we get back the moment of time that has passed by?

Since certainty subspace changes with bandwidth $\Delta \upsilon$, as illustrated in Figure 3.16, we see that as bandwidth $\Delta \upsilon$ decreases, a very large certainty subspace can be created within our universe, as depicted from Figure 3.16(a) to Figure 3.16(c).

High-resolution observation requires a shorter wavelength, but a shorter wavelength inherently has a broader bandwidth $\Delta \upsilon$ that creates a smaller certainty subspace, which can be used for high-resolution wave front reconstruction [21]. On the other hand, for a larger certainty subspace, it required a narrower bandwidth of $\Delta \upsilon$ that has a larger certainty subspace for exploitation, such as applied to side looking radar imaging [20]. In this we see that the size of the certainty subspace can be manipulated by the bandwidth $\Delta \upsilon$ as will be shown in the following.

A narrower bandwidth $\Delta \upsilon$ offers a huge certainty subspace that can be exploited for long-distance communication, in which I have found that the certainty subspace is in fact the coherence subspace, as I have discussed in the preceding. In other words, within a certainty subspace, it exhibits a point-pair certainty or coherent property among them, as illustrated in Figure 3.17. In other words, it has a high degree of certainty within a certainty subspace between points. This means that if a photonic particle has been started at point u_1, then it has a high degree of certainty that the particle is found at the next instance Δt at u_2, since distance is time within a temporal (t > 0) subspace.

For example, given any two arbitrary complex disturbances $u_1(r_1; t)$ and $u_2(r_2; t)$, as long as the separation between them is shorter than the radius Δr of the certainty subspace as given by:

$$d \leq c/(\Delta \upsilon) \tag{3.24}$$

the disturbances between $u_1(r_1; t)$ and $u_2(r_2; t)$ are "certainly" related (or mutually coherence). For this the "degree of certainty" (i.e., degree of coherence) between u_1 and u_2 can be determined by the following equation [15]:

$$\gamma_{12}(\Delta t) = \frac{\Gamma_{12}(\Delta t)}{\Gamma_{11}(0)\Gamma_{22}(0)} \tag{3.25}$$

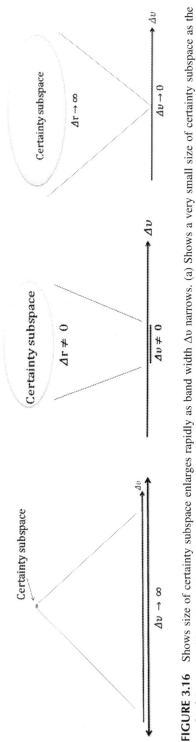
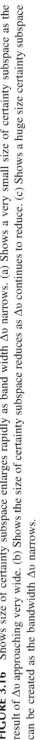

FIGURE 3.16 Shows size of certainty subspace enlarges rapidly as band width $\Delta\upsilon$ narrows. (a) Shows a very small size of certainty subspace as the result of $\Delta\upsilon$ approaching very wide. (b) Shows the size of certainty subspace reduces as $\Delta\upsilon$ continues to reduce. (c) Shows a huge size certainty subspace can be created as the bandwidth $\Delta\upsilon$ narrows.

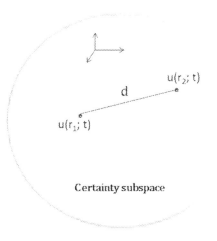

FIGURE 3.17 Shows mutual certainty within a certainty subspace. $u_1(r_1; t)$ and $u_2(r_2; t)$ represent two arbitrary disturbances separated at distance d.

where the mutual certainty (or mutual coherence) function between u_1 and u_2 can be written as:

$$\Gamma_{12}(\Delta t) = \lim_{T \to \infty} \frac{1}{T} \int_0^T u_1(t; r_1) u_2^*(t - \Delta t; r_1) dt \qquad (3.26)$$

Similarly, the respective "self-certainty" (or self-coherence) functions are respectively given by:

$$\Gamma_{11}(\Delta t) = \lim_{T \to \infty} \frac{1}{T} \int_0^T u_1(t; r_1) u_1^*(t - \Delta t; r_1) dt \qquad (3.27)$$

$$\Gamma_{22}(\Delta t) = \lim_{T \to \infty} \frac{1}{T} \int_0^T u_2(t; r_2) u_2^*(t - \Delta t; r_2) dt \qquad (3.28)$$

One of the interesting applications for a certainty principle must be the synthetic aperture radar imaging, as I have mentioned earlier, is shown in Figure 3.18. In this we see an aircraft carried a side looking synthetic radar system, shown in Figure 3.18(a), and emits a sequence of radar pulses scanned across the flight path of the terrain. The returned pulses are combined with local radar pulses, which are mutually coherent (i.e., high degree of certainty), to construct a recording format that can be used for imaging the terrain, for which a synthetic imagery is shown in Figure 3.18(b). Here we see a variety of scatters including city streets, wooded areas and farmlands, and lakes with some broken ice floes can also be identified on the right of this image. A microwave antenna has a very narrow carrier bandwidth (i.e., Δv) and its certainty radius (i.e., $d = c \cdot \Delta t$) or the coherence length can be easily

(a)   (b)

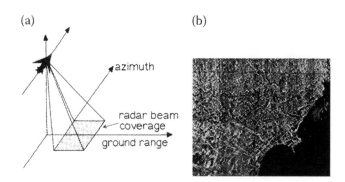

FIGURE 3.18 A side looking radar imaging within a certainty subspace: (a) shows a side looking radar scanning flight path and (b) shows an example of synthetic aperture radar imagery.

reached to hundreds of thousand feet. In other words, a very large certainty subspace for complex-amplitude imaging (or for communication) can be realized.

Finally, I would address again within the certainty unite $(\Delta p, \Delta r)$ [i.e., equivalently for $(\Delta E, \Delta t)$ and $(\Delta v, \Delta t)$ unit] can be mutually traded. But it is the trading of Δp for Δr (or ΔE for Δt and Δv for Δt) that is physically visible since time is not a physical substance but a forward constant dependent "variable" that we "cannot" manipulate. For this we see that the "section" of Δt that has been "used" cannot get it back. In other words, we can get back the same amount of Δt, but not the same moment of Δt that has been expensed. As I have shown earlier, everything within our universe has a price, an amount of energy ΔE and a section of time Δt. Aside from ΔE, we can physically change, but it is the moment of time Δt that had been expensed, preventing us to get it back. It is because that moment of Δt is the same moment of time of our temporal (t > 0) universe that has been passed. And this is the moment of time Δt, regardless of how small it is once it goes by, we cannot get it back. And this is the moment of Δt that Schrödinger's superposition principle wants to stop momentarily, but it could not.

3.9   REMARKS

I would remark that quantum scientists used amazing mathematical analyses added with their fantastic computer simulations to provide very convincing results. But mathematical analyses and computer animations are virtual and fictitious, and many of their animations are not physically real. For example, the instantaneous and simultaneous superimposing principle for quantum computing does not actually exist within our universe. One of the important aspects within our universe is that one cannot get something from nothing; there is always a price to pay, an amount of energy ΔE, and a section of time Δt. The important thing is that they are not free!

Since any science existed within our universe has time or temporal (t > 0), in which we see that any scientific hypothesis has to comply with a temporal (t > 0) condition within our universe, otherwise it may not be physically realizable science. Science is mathematics, but mathematics is not equal to science; any analytic solution has to be temporal (t > 0), otherwise it cannot be implemented within our universe. This includes all the laws, principles, and theories that have to be temporal (t > 0). Since time is a dependent variable that coexists with space, we have concluded that time is not an illusion but is real, as in contrast with most of the scientists who believe that time is an independent variable and some of them even believe that time is an illusion.

The uncertainty principle is one of most fascinating principles in the quantum theory, yet Heisenberg's principle was based on a diffraction-limited observation, and it is not due to the nature of time. But we have shown uncertainty increases naturally with time, as in contrast with Heisenberg's principle. I also introduced a certainty principle, from which a certainty subspace can be exploited for complex-amplitude communication. Yet the essence of this chapter is not how rigorous the mathematics is, but it is the physically realizable science we need since mathematics is not equal to science.

REFERENCES

1. A. Einstein, *Relativity, the Special and General Theory*, Crown Publishers, New York, 1961.
2. F. T. S. Yu, "From Relativity to Discovery of Temporal (t > 0) Universe", *Origin of Temporal (t > 0) Universe: Correcting with Relativity, Entropy, Communication and Quantum Mechanics*, Chapter 1, CRC Press, New York, 1–26, 2019. New York.
3. M. Bunge, *Causality: The Place of the Causal Principle in Modern Science*, Harvard University Press, Cambridge, 1959.
4. F. T. S. Yu, "The Fate of Schrodinger's Cat", *Asian Journal of Physics*, 28 (1): 63–70, 2019.
5. J. M. Knudsen, and P. Hjorth, *Elements of Newtonian Mechanics*, Springer Science & Business Media, 2012.
6. F. T. S. Yu, "Time: The Enigma of Space", *Asian Journal of Physics*, 26 (3): 143–158, 2017.
7. F. T. S. Yu, "What is "Wrong" with Current Theoretical Physicists?", *Advances in Quantum Communication and Information*, Edited by F. Bulnes, V. N. >. Stavrou, O. Morozov and A. V. Bourdine, Chapter 9, pp. 123–143, IntechOpen, London, 2020.
8. E. Parzen, *Stochastic Processes*, Holden Day, Inc., San Francisco, 1962.
9. Einstein Attacks Quantum Theory, *Scientist and Two Colleagues Find It Is Not 'Complete' Even Though 'Correct'*, The New York Times, April 28, 1935.
10. S. Hawking and R. Penrose, *The Nature of Space and Time*, Princeton University Press, New Jersey, 1996.
11. J. D. Kraus, *Electro-Magnetics*, McGraw-Hill Book Company, New York, 1953, p. 370.
12. F. T. S. Yu, *Optics and Information Theory*, Wiley-Interscience, New York, 1976.
13. E. Schrödinger, "An Undulatory Theory of the Mechanics of Atoms and Molecules", *Physical Review*, 28 (6): 1049, 1926.
14. W. Heisenberg, "Über den anschaulichen Inhalt der quantentheoretischen Kinematik und Mechanik", *Zeitschrift Für Physik*, 43 (3–4): 172, 1927.
15. F. T. S. Yu, *Introduction to Diffraction, Information Processing and Holography*, Chapter 10, MIT Press, Cambridge, Mass, 91–98, 1973.
16. D. F. Lawden, *The Mathematical Principles of Quantum Mechanics*, Methuen & Co Ltd., London, 1967.

17. Ludwig Boltzmann, "Über die Mechanische Bedeutung des Zweiten Hauptsatzes der Wärmetheorie", *Wiener Berichte*, 53: 195–220, 1866.
18. E. MacKinnon, "De Broglie's Thesis: A Critical Retrospective", *American Journal of Physics*, 44: 1047–1055, 1976.
19. D. Gabor, "A New Microscope Principle", *Nature*, 161: 777, 1948.
20. L. J. Cultrona, E. N. Leith, L. J. Porcello and W. E. Vivian, "On the application of Coherent Optical Processing Techniques to Synthetic-Aperture Radar", *Proceedings of the IEEE*, 54: 1026, 1966.
21. E. N. Leith and J. Upatniecks, "Reconstructed Wavefront and Communication Theory", *Journal of the Optical Society of America*, 52: 1123, 1962.

4 From Hamiltonian to Temporal (t > 0) Mechanics

Two of the most important discoveries in the twentieth century in modern science must be Einstein's relativity theory [1] and Schrödinger's quantum mechanics [2]; one deals with very large objects and the other deals with very small particles. They are connected by means of Heisenberg's uncertainty principle [3] and Boltzmann's entropy theory [4]. Yet, practically all the laws, principles, and theories of science were developed from an absolute empty space, and their solutions are all timeless (t = 0) or time independent. Our universe is a temporal (t > 0) space, for which a timeless (t = 0) solution cannot be "directly" implemented within our universe, because timeless and temporal are mutually exclusive. Although timeless laws and principles have been the foundation and cornerstone of our science, there are also scores of virtual solutions that are "not" physically realizable within our temporal (t > 0) space.

Yet, it is the major topic of the current state of science; as fictitious and virtual as mathematics is. Added with a very convincing computer simulation, fictitious science becomes "irrationally" real. As a scientist, I felt, in part, my obligation to point out where those fictitious solutions come from, since science is also mathematics.

Since Schrödinger's quantum mechanics is a legacy of Hamiltonian classical mechanics, I will first show that Hamiltonian was developed on a timeless (t = 0) platform, for which Schrödinger's quantum machine is also timeless (t = 0) this includes his quantum world as well as his fundamental principle of superposition. I will further show that where Schrödinger's superposition principle is timeless (t = 0), it is from the adaption of Bohr's quantum state energy $E = h\upsilon$, which is essentially a time-unlimited singularity that is approximated. I will also show a non-physical realizable wave function can be configured to become temporal (t > 0), since we knew a physically realizable wave function was supposed to be. And I will show that the superposition principle existed "if and only if" a timeless (t = 0) virtual mathematical subspace, but did not exist within our temporal (t > 0) space.

When dealing with quantum mechanics, it is unavoidable not to mention Schrödinger's cat, which is one of the most elusive cats in science, since Schrödinger disclosed the hypothesis in 1935. And the interesting part is that the paradox of Schrödinger's cat has been debated by a score of world-renowned scientists such as Einstein, Bohr, Schrödinger, and many others for over eight decades and it is still debating. Yet I will show Schrödinger'ss hypothesis is "not" a physically realizable hypothesis, for which his half-life cat should "not" have been used as a physically postulated hypothesis.

DOI: 10.1201/9781003271505-4

Since quantum communication and computing rely on a qubit information algorithm, I will show that qubit information logic is as elusive as Schrödinger's cat. It existed only within an empty space that has no price to pay, but does not exist within our temporal (t > 0) universe. This is always a price to pay, a section of time Δt and an amount of energy ΔE. Similarly I will show that a double-slit postulation is a timeless (t = 0) hypothesis that cannot exist within our temporal (t > 0) universe. What I mean is that double-slit postulation is another false hypothesis as Schrödinger's cat that had let us to believe superposition actually existed within our time-space.

In short, the art of quantum mechanics is all about physically realizable mechanics. In this we see that everything existed within our universe, no matter how small it is, and it has to be temporal (t > 0); otherwise it cannot exist within our universe.

4.1   HAMILTONIAN TO TEMPORAL (t > 0) QUANTUM MECHANICS

In modern physics, there are two important pillars of disciplines: It seems to me one is dealing with macro-scale objects of Einstein [1] and the other deals with micro-scale particles of Schrödinger [2]. Instead of speculating that micro- and macro-object behaves differently, they share a common denominator: temporal (t > 0) subspace. In other words, regardless of how small the particle is, it has to be temporal (t > 0), otherwise it cannot exist within our temporal (t > 0) universe.

As science progresses from Newtonian [5] to statistical mechanics [6], "time" has always been regarded as an "independent" variable with respect to substance or subspace. And this is precisely what modern physics has been used in the same timeless (t = 0) platform, for which they had treated time as an "independent" variable for centuries. Since Heisenberg [3] was one of the earlier starters in quantum mechanics, I have found his principle was derived on the same timeless (t = 0) platform, as depicted in Figure 4.1. And this is the "same" platform used in developing Hamiltonian classical mechanics [7]. Precisely this is the reason why Schrödinger's quantum mechanics is "timeless (t = 0)" since quantum mechanics is the legacy of Hamiltonian [8].

In view of Figure 4.1, we see that the background of the paradigm is a piece of paper that represents an empty subspace; it is "not" a physically realizable model since a particle and empty space are mutually excusive. Notice that the total energy

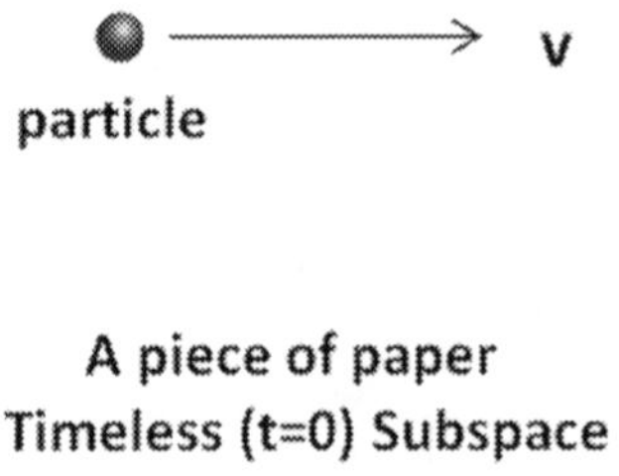

particle ●————————→ **v**

FIGURE 4.1   Shows a particle in motion within a time-less (t = 0) subspace. v is the velocity of the particle.

A piece of paper
Timeless (t=0) Subspace

of a "Hamiltonian particle" in motion is equal to its kinetic energy plus the particle's potential energy as given by [7]:

$$\mathcal{H} = p^2/(2m) + V \tag{4.1}$$

which is the well-known Hamiltonian equation, where p and m represent the particle's momentum and mass, respectively; and V is the particle's potential energy. Equivalently, the Hamiltonian equation can be written in the following form as applied for a "subatomic particle":

$$\mathcal{H} = -[h^2/(8\pi^2 m)]\nabla^2 + V \tag{4.2}$$

where h is Planck's constant, m and V are the mass and potential energy of the particle, and ∇^2 is a Laplacian operator:

$$\nabla^2 = \frac{\partial^2}{\partial xi \ \partial xj}$$

We note that Eq. (4.2) is the well-known "Hamiltonian Operator" in classical mechanics. Notice that, Hamiltonian is an energy zero-summed operator since it was developed from an empty paradigm as can be seen in figure:

Nevertheless, by virtue of "energy conservation", Hamiltonian equation can be written as:

$$\mathcal{H}\psi = \{-[h^2/(8\pi^2 m)]\nabla^2 + V\}\psi = E\psi \tag{4.3}$$

where ψ is the wave function that remains to be determined, E and V are the energy factor and potential energy that need to be incorporated within the equation. And this is precisely where Schrödinger's equation was derived from; by using the energy factor $E = h\upsilon$ (i.e., a quanta of light energy) adopted from Bohr's atomic model [9], Schrödinger equation can be written as [7]:

$$\frac{\partial^2\psi}{\partial x^2} + \frac{8\pi^2 m}{h^2}(E - V)\psi = 0 \tag{4.4}$$

In view of Schrödinger's equation, we see that it is essentially "identical" to the Hamiltonian equation, where ψ is the wave function to be determined, m is the mass of a photonic-particle (i.e., photon), E and V are the dynamic quantum state energy and potential energy of the particle, x is the spatial variable, and h is Planck's constant.

Since Schrödinger's equation is the "core" of quantum mechanics, but without Hamiltonian's mechanics it seems to me, would we "not" have quantum mechanics? The "fact" is that quantum mechanics is essentially "identical" to Hamiltonian mechanics. The major difference between them is that Schrödinger used a dynamic quantum energy $E = h\upsilon$ as adapted from a quantum leap energy of Bohr's hypothesis that changes from classical mechanics to quantum "leap" mechanics or quantum mechanics. In other words, Schrödinger used a package of wavelet quantum leap energy hυ equivalent to a particle (or photon) from "wave-particle dynamics" of de Broglie's hypothesis [10], although a photon is "not" actually a real particle (see Appendix B). Nevertheless, where the mass m for a photonic particle in the Schrödinger's equation remains to be "physically reconciled", after all science is a law of approximation? Furthermore without the adaptation of Bohr's quantum leap hυ, quantum physics would not have started. It seems to me that quantum leap energy E = hυ has played a "viable" role as transforming from Hamiltonian classical mechanics to quantum mechanics, which Schrödinger had done to his quantum theory.

Although Schrödinger's equation has given scores of viable solutions for practical applications, at the "same time" it has also produced a number of fictitious and irrational results that do not exist within our universe, such as his fundamental principle of superposition, the paradox of Schrödinger's cat [8], and others.

In view of Schrödinger's equation, as given by Eq. (4.4), we see that it is a timeless (t = 0) or time-independent equation. Since the equation is the "core" of Schrödinger's quantum mechanics, it needs a special mention. Let me stress the essence of energy factor E in the Hamiltonian equation. Does any solution that comes out from Schrödinger's equation undermines the wave equation ψ whether it is physically realizable or not? In other words, as the solution comes out from Schrödinger's equation if it is physically realizable or not depends upon the E factor that we introduce in Schrödinger's equation. Referring to the conventional Hamiltonian mechanics, if we let the energy factor E be a "constant" quantity that exists at time $t = t_0$, this is "exactly" the classical mechanics of Hamiltonian. This means that the Hamiltonian will take this value of E at $t = t_0$, and evaluates the wave function ψ, as has been given by [7]:

$$\psi = \psi_0 \exp[-i\ 2\pi\ E(t-t_0)/h) \tag{4.5}$$

This is the Hamiltonian wave equation, where ψ_0 is an arbitrary constant, h is Planck's constant, and a constant energy factor $E(t - t_0)$ occurs at $t = t_0$. Although the Hamiltonian wave equation is a time-variable function, it is "not" a time-limited solution. For this we see that it "cannot" be implemented within our temporal (t > 0) universe, since a time-unlimited solution cannot exist within our universe. This means that the wave solution ψ of Eq. (4.5) is "not" a physically realizable solution.

Then a question is being raised: why is the Hamiltonian wave solution time unlimited? The answer is trivial that Hamiltonian is mathematics, and his mechanics were developed in an empty timeless (t = 0) platform, as can be seen in Figure 4.1. Since it is the subspace that governs the mechanics, we see that particle-wave dynamics cannot exist within a timeless (t = 0) subspace. But Hamiltonian is mathematics and Hamilton himself is a theoretician, so he could have had implanted a particle-wave dynamic into a timeless (t = 0) subspace, although timeless (t = 0) subspace and a physical particle cannot coexist. Of this is precisely all the scientific laws, principles, and theories that were mostly developed on a piece or pieces of papers, since science is mathematics. This is by no means that timeless (t = 0) laws, principles, and theories were wrong, yet they were and "still" are the foundation and cornerstone of our science [11]. However, it is their direct implementation within our temporal ($t > 0$) universe and also added a score of their solutions are irrational and virtual as "pretending" to exist within our temporal ($t > 0$) subspace; for example, the superposition principle of quantum mechanics, the paradox of Schrödinger's cat, time travelling, and many others.

Nevertheless, as we refer to Figure 4.1, immediately we see that it is "not" a physically realizable model should be used in the first place. Secondly, even though we pretending that the particle in motion within can exist in an empty space, but a question is being asked; how can a particle-wave dynamic propagates within an empty space? Thirdly, even we assumed wave can be exited within an empty space, why it has to be time unlimited? From all these physical reasons, we see that time unlimited Hamiltonian wave equation of Eq. (4.5) is "not" a physically realizable solution since it only existed within a timeless ($t = 0$) virtual mathematical space. This is similar within a Newtonian space, where time has been treated as an "independent" variable.

Since Schrödinger's mechanics is the legacy of Hamiltonian mechanics, firstly we see that Schrödinger's quantum "mechanics" is a solution obtained from Hamiltonian's mechanics. Secondly, the reason why Schrödinger's quantum mechanics is timeless ($t = 0$) it is the same reason as Hamiltonian, because its subspace is empty. Nevertheless, the major differences between Schrödinger's mechanics and Hamiltonian mechanics must be the name's sake of "quantum", which comes from Bohr's atomic quantum leap $E = h\upsilon$, a quantum of light shown in Figure 4.2 that Schrödinger has used for the development of his mechanics. This is precisely

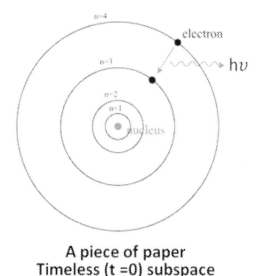

A piece of paper
Timeless (t =0) subspace

FIGURE 4.2 Shows Bohr's atomic model embedded in a timeless ($t = 0$) platform (i.e., a piece of paper).

Schrödinger's solution and is very similar to Hamiltonian's Eq. (4.3) as given by [7]:

$$\psi(t) = \psi_0 \exp[-i\ 2\pi\ \upsilon(t{-}t_0)/h] \tag{4.6}$$

This is the well-known Schrödinger wave equation, where ψ_0 is an arbitrary constant, the frequency of the quantum leap $h\upsilon$, and h is Planck's constant.

As anticipated, Schrödinger's wave equation is also a "time-unlimited" with "no" bandwidth. For the same reason as Hamiltonian, Schrödinger's wave equation is "not" a physically realizable solution that can be implemented within our temporal (t > 0) universe, since any physically realizable wave equation has to be "time and band limited". Yet many quantum scientists have been using this time-unlimited solution to pursue their dream for quantum supremacy computing [12] and communication [13], but "not" knowing the dream they are pursuing is "not" a physically realizable dream.

Since quantum mechanics is a "linear" system machine, similar to Hamiltonian mechanics. For a multi-quantum state energy atomic particle, the energy E factor to be applied in Schrödinger's equation is a "linear" combination of those quantum state energies as given by:

$$E = \sum h\upsilon_n, \quad n = 1, \quad 2, \quad \ldots \ N \tag{4.7}$$

where υ_n is the frequency for nth quantum leap and h is Planck's constant. Therefore, the overall wave equation is a linear combination of all the wave functions as given by:

$$\psi_N(t) = \sum \psi_{0n} \exp[-i\ 2\pi\ \upsilon_n(t{-}t_{0n})/h], \quad n = 1, \quad 2, \ldots N \tag{4.8}$$

In this we see that all the wave functions are "super-imposing" together. This is precisely the fundamental principle of superposition of Schrodinger. Yet, this is the principle that Einstein "opposed" the most, as he commented: "mathematics is correct, but incomplete" published in the *New York Times* newspaper in 1935 [14]. And it is also the fundamental principle that quantum computing scientists are depending on; the "simultaneous and instantaneous" superposition that quantum theory can offer to develop a quantum supremacy computer. I will show that the superposition is a timeless (t = 0) principle, and it does "not" exist within our universe.

Before I get started, it is interesting to show a hypothetical scenario of "superposition in life". If we assumed our life expectancy can last for about 500 years, then we would have a very good chance to coexist with Isaac Newton and possibly with Galileo Galilei somewhere in "time". Furthermore, if our universe is a "static" universe or timeless (t = 0), then we are also very likely to coexist with Galileo and Newton not only in "time" but superimposing with them everywhere in a timeless (t = 0) space. And this precisely what "simultaneous and instantaneous" superposition can do for us if our universe is a timeless (t = 0) subspace.

As we understood from the preceding illustration, we know that any empty (i.e., timeless) subspace cannot be found within our universe. And we have also learned that within our universe, every quantum leap hv has to be temporal (t > 0); that is, time and band are limited; otherwise it cannot exist within our universe.

In view of Eq. (4.7) and Eq. (4.8), we see that they are time "unlimited" wave functions, and it is trivial to see that all of those wave functions, $\psi_N(t)$, n = 1, 2 ... N, are superimposing together at all times. Similar to an example that I postulated earlier, if our life expectancy can be extended to 500 years, we would coexist with Einstein and may be with Newton somewhere in time, although 500 years life expectancy is time limited. But again, a time-unlimited wave function is "not" a physically real function since it cannot exist within our temporal (t > 0) universe.

In order to mitigate the temporal (t > 0) requirement or the causality condition of those wave functions, $\psi_N(t)$, we can "reconfigure" each of the wave functions to become temporal (t > 0). In other words, we can reconfigure each of the wave functions to "comply" with the temporal (t > 0) condition within our universe. For example, as illustrated in Figure 4.3, we see that each quantum leap $h\Delta v$ is represented by a "time-limited" wavelet.

By this It can be shown that "reconfigured" wave functions are approximated by:

$$\psi(t) = \sum \psi_{on} \exp[-\alpha_{on}(t - t_{on})^2]\cos(2\pi v_n t), \quad t > 0, \quad n = 1, \quad 2, \quad 3 \quad (4.9)$$

$$\psi(t) = 0, \quad t \leq 0 \qquad (4.10)$$

where t > 0 denotes an equation is subjected to a temporal (t > 0) condition in a positive time domain. In view of these equations, we see that the packages of quantum leaps are "likely" temporally separated. In this we see that all the wavelets are very "unlikely" to be "simultaneous and instantaneous" superposing together.

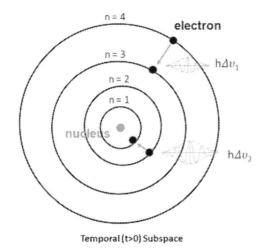

FIGURE 4.3 Shows a multi-quantum state atomic model embedded within a temporal subspace.

Once again, we have proven that Schrödinger's fundamental principle of super-position "fails" to exist within our temporal (t > 0) universe.

4.2   TIMELESS (T = 0) SPACE DOSE TO PARTICLES

On the other hand, if we take the preceding physically realizable wave functions of Eq. (4.9) and implement them within a timeless (t = 0) subspace, then it is trivial to see how a timeless (t = 0) subspace can do to all the wave-particle dynamics within a timeless (t = 0) subspace. Since within a timeless (t = 0) space it has no time and no dimension, all wave particles (i.e., package of wavelets) will collapse at t = 0, as can be seen in Figure 4.4.

Before we go on, I would say that the wave-particle duality is a "non-physical" reality assumption to the "equivalence" of a package wavelet of energy to a particle in motion, which is strictly from a statistical mechanics point of view, where the momentum of a particle $p = h/\lambda$ is conserved [7]. However, one should "not" treat a wave or a package of wavelet energy $h\Delta\upsilon$ as a particle or a particle as a wave. It is the package of wavelet energy "equivalent" to a particle dynamic (i.e., photon), but they are "not" equal [15]. Similar to Einstein's energy equation, mass is equivalent to energy and energy is equivalent to mass, but mass is not equal to energy and energy is not mass. For this a quantum of light $h\Delta\upsilon$ or a "photon" is a "virtual" particle. In this we see that a photon has a momentum $p = h/\lambda$ but no mass, although many quantum scientists regard a photon as a physical real particle.

In view of Figure 4.4 we see that within a timeless (t = 0) space, it has no time and no space; every particle exists anywhere within a timeless (t = 0) space but only exists at t = 0. This is precisely the "simultaneous and instantaneous" superposition

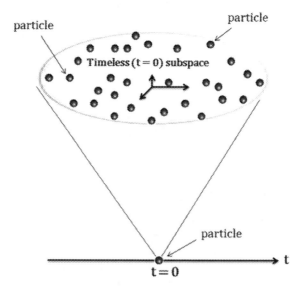

FIGURE 4.4   Shows all the particles within a timeless (t > 0) subspace; converges all the particles at t = 0.

Schrödinger's principle anticipated for. Since this is the fundamental principle that quantum scientists are aiming for, to build a quantum supremacy computer. And as well applied to quantum entanglement communication, unfortunately the "simultaneous and instantaneous" superposition does "not" exist within our universe. Of this we have shown, the superposition principle exists "if and only if" in a mathematical virtual timeless ($t = 0$) space and it cannot exist within our temporal ($t > 0$) universe.

The reason that the superposition principle "fails" to exist is coming from a non-physical realizable paradigm used in the analysis, which can be traced back to the development of Hamiltonian mechanics, since quantum mechanics is an extension of Hamiltonian mechanics. I have found It is the background subspace (i.e., a piece of paper) used in quantum mechanical analysis, since the background represents an "inadvertently" empty timeless ($t = 0$) subspace, where a photonic particle in motion was embedded. And it is also that piece of paper that Bohr's atomic model was used. Adding his quantum state energy $h\upsilon$ is not a time-limited physical reality.

The substance and emptiness are mutually excluded; it is the subspace that governs the behavior of each wave function $\psi_n(t)$. In this timeless ($t = 0$) subspace, we have shown all the wave functions $\psi_N(t)$, regardless of time limited or time unlimited, they collapse all together at $t = 0$. In other words, all the quantum state wavelets superimposed at a "singularity" $t = 0$. This is the reason superposed quantum state energies can be found anywhere and everywhere within a virtual mathematical timeless ($t = 0$) space, since a timeless ($t = 0$) space has no distance.

From this illustration we have shown once again that it is not how rigorous the mathematics is, it is the physically realizable paradigm that determines its analytical solution is physically realizable or not. For this we see that the wave functions obtained from Schrödinger's equation are as virtual as mathematics is, because Schrödinger's quantum mechanics were developed on an empty subspace platform, the same platform as Hamiltonian classical mechanics.

4.3 SCHRÖDINGER'S CAT

When we are dealing with quantum mechanics, it is inevitable to mention Schrödinger's cat since it is one of the most elusive cats in science since Schrödinger disclosed it in 1935 at a Copenhagen forum. Since then, his half-life cat has intrigued a score of scientists and has been debated by Einstein, Bohr, Schrödinger, and many others as soon as Schrödinger disclosed his hypothesis. And the debates have persisted for over eight decades, and still debating. For example, I may quote one of the late Richard Feynman quotations as: "After you have leaned quantum mechanics, you really 'do not' understand quantum mechanics ...".

It is however not the art of the Schrödinger's half-life cat; it is the paradox that quantum scientists have treated it as a physical "real paradox". In other words, many scientists believed the paradox of Schrödinger's cat actually existed within our universe, without any hesitation. Or literally "accepted" superposition is a physical reality, although fictitious and irrational solutions have emerged; it seems like looking into the Alice wonderland. In order to justify some of their beliefs, some quantum scientists even come up with their beliefs; particles behave weird within a microenvironment in contrast within a macro-space. Yet some of their potential applications, such as quantum computing and quantum entanglement communication, are in fact in a macro subspace environment. Nevertheless, I have found many of those micro behaviors have "not" existed within our universe; and the paradox of Schrödinger's cat is one of them, as I shall discuss briefly in the following.

Let us start with Schrödinger's box, as shown in Figure 4.5; inside the box we have equipped a bottle of poison gas and a device (i.e., a hammer) to break the bottle, triggered by the decaying of a radio-active particle, to kill the cat. Since the box is assumed totally opaque of which no one knows that the cat will be killed or not, as imposed by Schrödinger's superposition principle until we open his box.

As we investigated Schrödinger'ss hypothesis of Figure 4.5, immediately we see that it is "not" a physically realizable postulation after all, since within the box it has a timeless (t = 0) or time-independent radioactive particle in it. As we know that any particle within a temporal (t > 0) subspace has to be a temporal (t > 0) particle or has time with it, otherwise the proposed radioactive particle cannot exist within Schrödinger's temporal (t > 0) box. It is, therefore, the paradox of Schrödinger's cat that is "not" a physically realizable hypothesis and we should "not" have treated Schrödinger's cat as a physically real paradox.

Since every problem has multiple solutions, I can change the scenarios of Schrödinger's box a little bit, such as allow a small group of individuals take turns to open the box. After each observation, close the box before passing on to the next observer. My question is how many times the superposition has to collapse. With all those apparent contradicted logics, we see that Schrödinger'ss cat is "not" a paradox after all! And the root of a timeless (t = 0) superposition principle as based on Bohr's quantum leap hυ, represents a time "unlimited" radiator, which is a singularly approximated wave solution. For this we should "not" have treated quantum leap hυ as a physical real radiator, since any quantum leap has to be time and band limited within our universe.

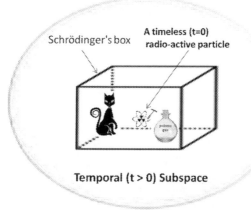

FIGURE 4.5 Paradox of Schrodinger's cat: Inside the box, we equipped a bottle of poison gas and a device (i.e., hammer) to break the bottle, triggered by the decaying of a radio-active particle, to kill the cat.

Finally, I would address that all the laws, principles, theories, and paradoxes were made to be broken, revised, and eradicated, and it is not that they were all approximated, because they all change with time or are temporal (t > 0). Yet, without approximated science, then there would be no science. In this we have shown that a simple hypothetical paradox takes decades to resolve! And this is the nature of quantum mechanics; it is all about temporal (t > 0) subspace.

4.4 NATURE OF Δ T

Our universe was assumed to be created with a huge energy explosion with time situated within a "non-empty" space. Every subspace "no" matter how small is created by an amount of energy ΔE and a section of time Δt for which every subspace is temporal (t > 0) (i.e., exists with time) (see Appendix A).

In view of modern science, there is a set of simple, yet elegant laws and principles that are profoundly associated with a unit of (ΔE, Δt). The objective of this section is to explore the relationship between these laws and principles as related with the unit of (ΔE, Δt). Since time is a dependent-forward variable that moves at a constant speed, we see that Δt is one of the most "esoteric" variables that exists within our universe. We will show that once a moment of Δt is used, we "cannot" get it back although ΔE and Δt can be traded. In this I will show that there it is a physical limit for Δt to approaching "none" (i.e., Δt ⟶ 0), that "prevents" us to reach; even though we have the all the price to pay. And this must be the nature of Δt?

Nevertheless, there is a set of "simple and elegant" laws and principles that are profoundly associated with a section of time Δt. These are laws, principles, and theories, such as entropy of Boltzmann [4], information of Shannon [16], uncertainty of Heisenberg [3], relativity of Einstein [1], and temporal (t > 0) space of Yu [6]. Each of them is associated with a section of time Δt that changes naturally with time. But the evidence tells us science has to be temporal (t > 0) and dynamics, which cannot be "static" or timeless (t = 0). In other words, if there is no time, then there has no science. Nevertheless, science is a law of "approximation", as in contrast with mathematics, which is an axiom of "certainty", of which I state these laws and principles "approximately" as follows:

Law of entropy; entropy within an enclosed subspace increases naturally "with time" or remains constant.

Theory of information; the higher the amount the information, the more uncertain the information is.

Principle of uncertainty; uncertainty of an isolated particle increases naturally "with time".

Theory of special relativity; when a subspace moves faster "relatively" than the other subspace, there is a "relativistic" time speed between them, although time speed within the subspaces remains the same. (Notice that this theory will be rebutted in Chapter 5 since it is not a physically realizable theory.)

Nature of the universe; every isolated subspace was created by the amount of energy ΔE and a section of time Δt within a temporal (t > 0) stochastic subspace changes naturally with time.

Nevertheless, it is easier to facilitate these laws and principles in mathematical forms, since mathematics is a "language", as respectively given by:

$$S = -k \ln p \tag{4.11}$$

$$I = -\log_2 p \tag{4.12}$$

$$\Delta E \cdot \Delta t \geq h \tag{4.13}$$

$$\Delta t' = \frac{\Delta t}{\sqrt{1 - v^2/c^2}} \tag{4.14}$$

$$\Delta E \ \Delta t \geq [(1/2) \ \Delta m \ c^2] \ \Delta t \tag{4.15}$$

where S, I, and U are entropy, information, and universe, respectively; k is the Boltzmann's constant; h is Planck's constant; p is the probability; Δt is a section of time; $\Delta t'$ is the dilated section of time; v is the velocity; m is the mass; and c is the speed of light. $k = 1.38 \times 10^{-16}$ ergs per degree centigrade and $h = 6.624 \times 10^{-27}$ erg-second.

In this we see that our universe was created by a means of a "huge" amount of energy ΔE and a "long" section of time Δt. And Δt is "still" extending rapidly since the boundary of our universe is still expanding at the speed of light [17].

In view of these laws and principles, they must be the most "elegant and simple" science equations that exist today. In this equations either are attached or associated with a section of time Δt, except Eq. (4.12) since information theory is mathematics. But as soon as information is recognized as related to entropy, information is equivalent to an amount of entropy; this makes an amount of information a physical quantity that is acceptable in science. For this we will show that a section of Δt will be associated with the theory of information; otherwise information will be very difficult to apply in science. Since Δt is coexisted with ΔE, we will further see that every bit of information takes an amount of energy ΔE and a section of time Δt to transmit, to create, to process, to store, to process, and to "tangle".

As we went back from Eq. (4.11) to Eq. (4.15) we see that they are all point-singularity approximated; otherwise it will be very difficult to write in simple mathematical forms. As the laws and principles are stated, there are all associated with time, by which they are all space-time laws and principles, since time is space and space is time within our temporal (t > 0) universe. In short, they are all connected to a unit of (Δt, ΔE), which is the basic building blocks of our universe. For this I envision that every existence within our universe has a beginning and has an end. But it is time; it has "no" beginning and has "no" end!

Since our temporal (t > 0) universe was created based on a commonly accepted Big Bang Theory [17], we see that our universe is a temporal (t > 0) dynamic "stochastic" subspace [18]. The boundary of our universe increases at the speed of light, and we see that every subspace within our universe is a "non-empty" temporal (t > 0) stochastic subspace. By the way, one- or two-dimensional subspaces "cannot" exist within our universe, since one- or two-dimensional subspaces are volumeless. For this any independent Euclidian subspace "cannot" be simply applied to describe a temporal (t > 0) subspace. Because all the dimensional coordinates (e.g., x. y z coordinates) of a temporal space are all "interdependent" with time, where time is a forward variable with respect to the subspace. In other words, every substance, no matter how small it is, has to have time and is temporal (t > 0).

In view of the time dilation of Einstein's relativity of Eq. (4.14) and Heisenberg's uncertainty principle of Eq. (4.13), we see that they are associated with a section of time Δt, which represents a "temporal (t > 0)" subspace, as given by:

$$\Delta r = c \; \Delta t \qquad (4.16)$$

where r is the radius of a spherical subspace and c is the velocity of light. In this we see that subspace enlarges rapidly as Δt increases is given by:

$$V = (\tfrac{3}{4}) \; \pi \; (c \; \Delta t)^3 \qquad (4.17)$$

This shows precisely our universe is expanding with a section of time Δt. Since ΔE is a physical quantity equivalent to a subspace that "cannot" be empty and coexisted with Δt, then every unit (ΔE, Δt) is a temporal (t > 0) subspace, in which we see that time and space "cannot" be separated. In other words, time and space are "interdependent", although ΔE is a physical quantity but Δt is an invisible "real" variable.

4.5 ENTROPY AND INFORMATION

As we look back at Boltzmann's entropy Eq. (4.11), we see that it is a typical timeless (t = 0) point-singularity approximated equation. But the law described entropy increases with "time", and implies that entropy is associated with a section of time Δt, although it is "not" shown in the equation. Nevertheless, the law of entropy is essentially identical to the law of information, as can be seen by the logarithmic expressions of Eq. (4.11) and Eq. (4.12), for which we have the following relationship as given by [19]:

$$S = k \; I \; \ln 2 \qquad (4.18)$$

where I is an amount of information in "bit" and k is the Boltzmann's constant. In this we see that "every bit" of information is equal to an amount of entropy ΔS as given by:

$$\Delta S = k \ \ln \ 2 \ \text{ per bit of information} \tag{4.19}$$

Although an amount of information can be "traded" for a quantity of entropy, entropy is a "cost" in energy "equivalent" to an amount of information, but "not" the "actual" information. In other words, it is a "necessary cost" of an amount of entropy to pay for an amount of information in bits. For example, if an amount of entropy ΔS is equivalent to 1,000 bits of information of a specific book, then how many books have the same 1,000 bits or how many different items also have 1,000 bits? Similarly, an amount of information in bits is not given as the actual information, but it is a "necessary cost" but "not sufficient" to obtain the precise information. In this we see that the amount of entropy ΔS is a "necessary cost" needed to obtain an equivalent amount of information in bits.

Since entropy is a "physical quantity" similar to energy, as given by:

$$\Delta S = \Delta E/T = h\Delta \upsilon/T \tag{4.20}$$

where $\Delta E = h\Delta\upsilon$ is the quantum leap energy and $T = C + 273$ is the absolute temperature in Kelvin and C is the temperature in degrees Celsius. In this we see that higher thermal noise requires higher energy to transmit a of bit information:

$$\Delta E = T \ k \ \ln \ 2 \tag{4.21}$$

Thus, we see that an amount of entropy is equivalent to an amount of information, but it is "not" the information. But an amount of information is equivalent to an amount of entropy that makes information a very "viable" physical quantity that can be applied in science. In this we see that information and entropy can be simply traded as given by:

$$\Delta S \Leftrightarrow \Delta I \tag{4.22}$$

Nevertheless, we have shown that either information or entropy has to be a temporal (t > 0) or time-dependent law, as given by, respectively:

$$I(t) = -\log_2 \ p(t), \quad t > 0 \tag{4.23}$$

$$S(t) = -k \ \ln \ p(t), \quad t > 0 \tag{4.24}$$

where k is the Boltzmann's constant. In this we see that either information or entropy "increases" with time, and (t > 0) denotes imposition by temporal (t > 0) constraint. The amount of entropy for I(t) bits of information can be written as:

$$S(t) = k \ I(t) \ \ln 2, \quad t > 0 \tag{4.25}$$

where I(t) is in bits and k is Boltzmann's constant. In view of the preceding equation, it shows that entropy increases as the amount information increases. In this we see that "every bit" of information ΔI takes an amount of energy ΔE and a section of time Δt to "create" or to transmit as given by:

$$\Delta I \sim \Delta E \; \Delta t = h, \quad \text{per bit of information} \qquad (4.26)$$

Since "every bit" of information is equivalent to an amount of entropy ΔS:

$$\Delta S = k \; \ln \; 2, \quad \text{per bit of information} \qquad (4.27)$$

Thus, every quantity of entropy ΔS is "equivalently" equal to an amount of energy ΔE and a section of time Δt to produce as shown by:

$$\Delta S = \Delta E / T \qquad (4.28)$$

where $T = C + 273$ is the absolute thermal noise temperature in Kelvin, C is the temperature in degrees Celsius, and h is Planck's constant. Since ΔE "coexisted" with Δt, it is reasonable to say that every ΔS is also associated with a section of time Δt as given by:

$$\Delta S \sim E \; \Delta t / T = h / T, \quad \text{per bit of information} \qquad (4.29)$$

In this we see that information is connected with the law of uncertainty, where "every bit" of information is profoundly associated with ΔE and Δt.

Since every subspace within our universe is created by an amount of energy ΔE and a section of time Δt, we see that Boltzmann's entropy, Shannon's information, Heisenberg's uncertainty, and Einstein's relativity has a profound association with a section of Δt and of ΔE since they coexist. In other words, all the laws, principles, and theories as well as the paradoxes have to comply with the "coexistence" of ΔE and Δt; otherwise those laws and principles cannot be guaranteed to exist within our universe.

Nevertheless, increasing entropy is regarded as a "degradation" of energy by Kelvin [19] although entropy was originated by Clausius [19]. But he might have intended it to be used as a "negative" of entropy (i.e., neg-entropy). In this we see that as entropy or the amount of information increases means that there is "energy degradation". This is also means that entropy or amount of information "degrades with time". Let me stress again: "energy degradation" within our universe is due to a boundary expansion of our universe at the speed of light [17]. For this I see it; that entropy increases with time is "no longer" a myth, as most scientists believe it is.

Since all the laws and principles are attached with a price tag of $(\Delta E, \Delta t)$, it is the $\Delta t \longrightarrow 0$ that "cannot" be reached, even though we assumed having all the energy of ΔE to pay for! This is precisely the "physical" limit of a temporal ($t > 0$) subspace, by which the "instantaneous" moment of time (i.e., $t = 0$) can be approached but can "never" be able to attend, regardless how much energy ΔE we willing to pay for. And this is the nature of Δt!

4.6   UNCERTAINTY AND INFORMATION

Every substance or subspace has a piece of information that includes all the elementary particles, basic building blocks of the subspaces, atoms, papers, our planet, solar system, galaxy, and even our universe! In other words, the universe is flooded with information (i.e., spatial and temporal), or information fills up the whole universe. Strictly speaking, when one is dealing with the origin of the universe, the aspect of information has never been absence. Then, one would ask: What would be the amount of information, aside from the needed energy ΔE, that is required to create a specific substance? Or equivalently, what would be the "cost" of entropy to create it? To answer this question, let me start with the law of uncertainty, in an equivalent form, as given by:

$$\Delta v \; \Delta t = 1 \tag{4.30}$$

where Δv is the bandwidth. In this there exists a profound relationship of an "information cell" [20], as illustrated in Figure 4.6. In this we see that the shape of (Δv, Δt) or equivalently (ΔE, Δt) can be "mutually" exchanged. Since every bit of information can be efficiently transmitted, if and only if it is transmitting within the constraint of the uncertainty principle (i.e., $\Delta v \cdot \Delta t \geq 1$). This relationship implies that the signal bandwidth should be either equal or smaller than the system bandwidth (i.e., $1/\Delta t \leq \Delta v$). In this we see that Δt and Δv can be "traded".

It is however the unit region but not the shape of the information cell that determines the limit; as illustrated in Figure 4.6, we see that within each unit cell, (Δv,

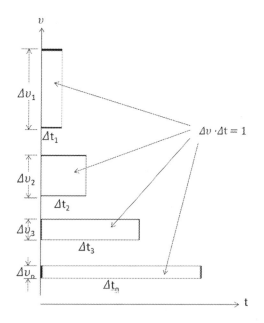

FIGURE 4.6   Shows various (Δv, Δt) information cells; where Δv_n and Δt_n are the bandwidths and time-limited sections, and $v_1 > v_2 > v_3 > \dots > v_n$ are the frequencies.

Δt) [or equivalently (ΔE, Δt)] can be mutually traded. But it is from Δυ to Δt or from ΔE to Δt, since Δυ and ΔE are physical quantities. For this we see that once a section of Δt is "used", we "cannot" get back the same moment of Δt, although we can create the same section of Δt, since time is a forward-dependent variable.

Nevertheless, there are basically two types of information transmission; one is limited by the uncertainty principle and the other is constrained within the "certainty subspace". And the boundary between these two regimes is given by $\Delta\upsilon \cdot \Delta t = 1$ (or $\Delta E \cdot \Delta t = h$), as I called this limit a quantum unit [21]. In this we see that Δυ can be traded for Δt. But under an uncertainty regime, information is carried by means of intensity (i.e., amplitude square) variation. Yet, information can also be transmitted within the certainty regime, such as applied to complex-amplitude communication [22,23]. As limited by the law of uncertainty, a quantum unit subspace QLS, for (ΔE, Δt) and (Δυ, Δt), is shown in Figure 4.7 for reference.

Since every subspace within our universe is a temporal (t > 0) subspace, the radius of any subspace can be described by a time-dependent variable as given by:

$$r = c \cdot \Delta t \tag{4.31}$$

where c is the speed of light and Δt represents a section of time. In this we see that the size of the subspace enlarges rapidly as Δt increases as given by:

$$V = (¾) \, \pi \, (c \, \Delta t)^3 \tag{4.32}$$

Since the carrier bandwidth Δυ and time resolution Δt are exchangeable, we see that the size of the QLS enlarges as the carrier bandwidth Δυ decreases. In other words, a narrower carrier bandwidth Δυ has the advantage of having a larger quantum-limited subspace for complex-amplitude communication, as depicted in Figure 4.8.

In this we see that it is possible to create a temporal (t > 0) subspace within a temporal (t > 0) space (i.e., our universe) for communication. We stress that it is "not" possible to create any time-independent or timeless (t = 0) subspace within our temporal universe, since timeless (t = 0) or time independent "cannot" exist within a temporal universe. And this timeless (t = 0) or the "instantaneous limit" (or the causal condition) is the fact of physical limit (i.e., $\Delta t \longrightarrow 0$) within our universe. This limit can only be approached with a huge amount of energy ΔE, but can we "never" be able to reach it?

Furthermore, let me note that a timeless (t = 0) or time-independent subspace is "not" an "inaccessible" space as some scientists claimed; since inaccessible implies

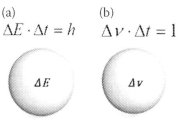

(a)   (b)

$\Delta E \cdot \Delta t = h$ $\Delta v \cdot \Delta t = 1$

ΔE   Δv

FIGURE 4.7   Shows a set of quantum limited subspaces (QLS).

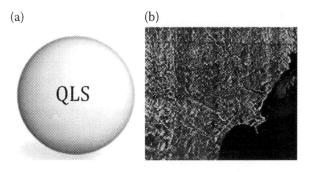

FIGURE 4.8 A "very large" quantum limited subspace as depicted in (a) can be realized in practice within our temporal (t > 0) space. For example, applied synthetic aperture radar imaging is shown in (b).

it existed within our universe. Nevertheless, one of the apparent aspects for using a large quantum-limited subspace is for a complex information transmission. For example, as applied to a complex wave front construction (i.e., holographic recording) [23], complex-match filter synthesis [24], as well apply to synthetic aperture radar imaging [22]. But there is an apparent price paid for using a "wider" section of time Δt; but it "deviates" farther away from real-time transmission.

4.7 RELIABLE COMMUNICATION

One of the important aspects for information transmission is that "reliable" information can be transmitted, such that information can be reached to the receiver with a "high degree of certainty". Let me take two key equations from the information theory: "mutual information" transmission through a "passive additive noise channel" is given by [19]:

$$I(A; B) = H(A)-H(A/B) \tag{4.33}$$

and

$$I(A; B) = H(B)-H(B/A) \tag{4.34}$$

where $H(A)$ is the information provided by the sender, $H(A/B)$ is the information loss (or equivocation) through transmission due to noise, $H(B)$ is information received by the receiver, and $H(B/A)$ is a noise entropy of channel.

However, there is a basic distinction between these two equations; one is for "reliable" information transmission and the other is for "retrievable" information. Although both equations represent the mutual information transmission between sender and receiver, their objectives are rather different. Example, using Eq. (4.33) is purposely used for "reliable information transmission" in which the transmitted information has a high degree of "certainty" to reach the receiver. While Eq. (4.34) is purposely used to "retrieve information" from "unreliable" information" by the receiver. For this we see that for "reliable" information transmission, one can simply increase the signal to noise ratio at the transmitting end. While for "unreliable" information, transmission is to extract information from ambiguous information. In other words, one is to be sure information will be reached to the receiver "before" information is transmitted, and the other is to retrieve the information "after" information has been received.

In communication, basically it has two basic orientations: one by Norbert Wiener [25,26] and the other by Claude Shannon [16]. But there is a major distinction between them; Wiener's communication strategy is that if the information is corrupted through transmission, it may be recovered at the receiving end, but with a "cost" mostly at the receiving end. While Shannon's communication strategy carries a step further by encoding the information before it is transmitted, such that information can be "reliably" transmitted, also with a "cost" mostly at the transmitting end. In view of the Wiener and Shannon information transmission orientations, a mutual information transfer of Eq. (4.33) is kind of a Shannon type, while Eq. (4.34) is kind of a Wiener type. In this we see that "reliable" information transmission is basically controlled by the sender; it is to "minimize" the noise entropy $H(A/B)$ (or equivocation) of the channel, as shown by:

$$I(A; B) \approx H(A) \tag{4.35}$$

One simple way to do it is by increasing the signal to noise ratio, with a "cost" of higher signal energy (i.e., ΔE).

On the other hand, to recover the transmitted information is to "maximize" $H(B/A)$ (the channel noise). Since the entropy $H(B)$ at the receiving end is "larger" than the entropy at the sending end, that is, $H(B) > H(A)$, we have:

$$I(A; B) = H(B) - H(B/A) \approx H(A) \tag{4.36}$$

Equation (4.35) essentially shows us that information can be "recovered" after being received, again with a price; ΔE and Δt. In view of these strategies, we see that the cost paid for using the Wiener type for information transmission is "much higher" than the Shannon type; aside from the cost of higher energy of ΔE, it needs an extra amount of time Δt for "post processing". Thus, we see that the Wiener communication strategy is effective for a "none cooperating" sender; for example, such as applied to radar detection and others. On the other hand, Shannon type provides a more reliable information transmission by simply increasing the signal to noise so that every bit of information can be "reliably transmitted" to the receiver.

Therefore, we see that quantum entanglement communication [13] is basically using the Wiener communication strategy. The price to pay will be "much higher and very inefficient" since it required intensive post-processing. And it is "illogical" to require the received signal to be "more equivocal" (i.e., uncertain); the better the information recovery, it can be received at the receiving end. In this, quantum entanglement communication is designed for extracting information as the Weiner communication. However, it is "not" the purpose for reliable information-transmission of Shannon.

4.8   TIMELESS (T = 0) QUBIT INFORMATION

Qubit information transmission is basically exploiting Wiener's communication strategy for the purpose of qubit transmission [12]. For this the receiver would anticipate a more ambiguous digital signal (e.g., either 0 or 1) from the transmitter. From this it has treated the receiving end entropy H(B) as a source entropy H(A) to determine the intended signal was sent. Since the signal was originated by the sender, by maximizing the entropy H(B) under a noiseless condition, the receiver can interpret the received signal (e.g., 0 or1) as equal to a qubit information. And this is precisely the qubit information principle that currently is using for quantum communication and computing [12,13]. In other words, the receiver is guessing the signal is either I or 0 after the signal has reached the receiving end. This is similar to receiving an envelope before opening it.

Nevertheless, it is not the question of the qubit transmission principle, but it is the physically realizable qubit information within our temporal (t > 0) universe. Since everything within our universe has a price to pay, qubit information transmission cannot be the exception. Firstly, quantum communication relies on the fundamental principle of superposition, but we have shown that the superposition principle cannot exist within our temporal (t > 0) universe. Then there is no sense in talking about all the possible capabilities that qubit information can offer.

Nevertheless, let me assume a quantum communication channel is situated within an empty space paradigm, shown in Figure 4.9; then we see that the quantum information channel is situated within in a timeless (t = 0) subspace and it has no time and no distance. From this we see that a binary source of A = {0, 1} is capable of transmitting 0 and 1 instantaneously and simultaneously within an empty space. This is precisely the same subspace platform that Schrödinger's fundamental principle of superposition derived from. From this we see that qubit information

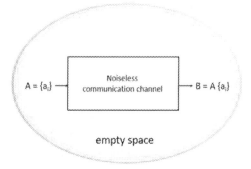

FIGURE 4.9   Shows a conventional noiseless communication channel is embedded within an empty space. But it is not a physically realizable paradigm since substance and emptiness cannot coexist.

transmission can only exist within an empty space, which is not a physically realizable information hypothesis that can be implemented within our time-space, since it has no time to represent a transmitting signal. The fact is that every temporal information (i.e., 0 or 1) needs a section of time (i.e., Δt) to present a time-signal, for a signal without a section of time (i.e., $\Delta t = 0$) represents a no-time signal. In other words, it has no carrier to transmit information within our temporal $(t > 0)$ universe, since qubit information is timeless $(t = 0)$.

It is not a physically realizable paradigm; let us find out how a qubit information channel works as depicted by a block box diagram shown in Figure 4.10, which is a timeless $(t = 0)$ noise-free channel. Where A = {0, 1} represents an input binary source, H(A) = 1 bit is the input entropy, B{qubit} is output quantum bit, and H(B) = qubit is the output entropy. Since the quantum qubit information transmission has treated the input binary source A = {0, 1} and the output ensemble is qubit B = {qubit}, such that at the receiving end information can be presented in quantum bit (i.e., qubit). But the qubit channel is embedded within a timeless $(t = 0)$ subspace; it has no noise and no time, and we see that it has no channel noise entropy [i.e., H(A/B) = 0]. From this, mutual information of the qubit channel can be written as:

$$I(A; B) = H(B) = H(A) \qquad (4.37)$$

where the output end entropy H(B) is equal to the input entropy H(A) [i.e., H(B) = H (A)]. Thus, we see that the intended sent signal is either 1 or 0, but not both, and is received at the receiving end. This is equivalent to recovering the intended input signal that was corrupted within a noisy channel of Wiener's information transmission.

Since quantum information is dependent on Schrodinger's superposition principle, such that binary transmission by which 0 and 1 can be transmitted instantaneously and simultaneously, this presents a quantum bit or a qubit to determine the input source encoding of either 1 or 0. But the quantum information channel is assumed within an empty space paradigm, and we see that the operation is instantaneous and simultaneous but only exists within a timeless $(t = 0)$ space. Since qubit information is the anchor principle for quantum computing and communication, unfortunately qubit information cannot exist within our temporal $(t > 0)$ universe.

A similar scenario to qubit information transmission is the paradox of Schrodinger's cat, where a received signal is dependent upon on observation. For example, the observer (i.e., the receiver) did not know the cat within Schrödinger's box was either alive or dead until the observer opened up the box. In this we see that the observer confirms the outcome after the observation. But the physical fact is that the cat is alive or dead

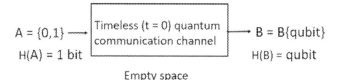

$$A = \{0,1\} \longrightarrow \boxed{\begin{array}{l} \text{Timeless } (t = 0) \text{ quantum} \\ \text{communication channel} \end{array}} \longrightarrow B = B\{qubit\}$$

H{A} = 1 bit   H(B) = qubit

Empty space

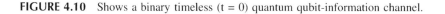

FIGURE 4.10   Shows a binary timeless $(t = 0)$ quantum qubit-information channel.

and has been determined before the observer opens up Schrödinger's box. Similarly, we never know a boiled egg is either hard or soft-boiled until we crack it open. But a hard- or soft-boiled egg had been determined before we crack the egg.

Although the paradox of Schrödinger's cat has been debated since the disclosure of the hypothesis in 1935, it seems to me that no one has found the reason where the paradox comes from until the recent discovery of the temporal (t > 0) universe [17,27]. From is I have shown that the paradox came from an empty subspace (i.e., a piece commonly used paper) where Schrödinger's equation derived from. From this I have shown that his fundamental principle of superposition is timeless (t = 0), and fails to exist within our universe.

On the other hand, if the qubit information channel is situated within a temporal (t > 0) subspace, as shown in Figure 4.11, then it responds to a supposed qubit channel that is subjected to the boundary condition within our time-space. In this we see that simultaneous and instantaneous superposition of binary digital transmission (i.e., 0, 1) fails to exist. From this we see that output entropy H(B) at the transmitted end cannot be treated as a qubit information since the superposition principle does not hold within our temporal (t > 0) space. Thus, we see that the output ensemble is B = {0, 1} instead of B = {qubit}, which is identical to a conventional noisy binary channel.

Let me further hypothesize a noiseless communication scenario, as diagramed in Figure 4.12, in which we assumed a "1" message was inserted within an envelope

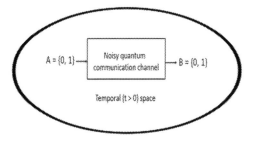

FIGURE 4.11   Shows a binary noisy quantum communication channel embedded within a temporal (t > 0) space. For this, output entropy is always larger than the input entropy; that is H(B) > H(A). Note: If it is a noise-free channel then H(B) = H(A), but a noiseless channel is not a physically realizable channel.

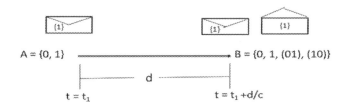

FIGURE 4.12   Shows an idea noiseless communication channel. In this a binary message was sent within an enclosed envelope from A to B. Receiver B does not know either 0 or 1 until the receiver opens up the envelope. c is the speed of light.

sent through a noiseless (i.e., without tempering) channel to a receiver at B. Since the receiver does not know the message within the envelope until he opens it, this is similar to the half-life cat of Schrödinger, until the observer opens Schrodinger's box.

Since the message within the envelope had been determined before it was sent, we see that it is not the receiver's consciousness that changes the outcome of the message; similarly, the life of Schrödinger's cat, which had been determined before the observer opens Schrödinger's box. Nevertheless, the transmission from sender A to receiver B is still not instantaneous, since the enclosed message takes a section of time Δt (i.e., $\Delta t = d/c$) to transmit and also an amount of energy ΔE associated with transmission to make it happen, and again it is not free. Furthermore, to represent a binary signal, it also needs another section of time $\Delta t'$. Notice that no matter how small $\Delta t'$ is, it has no signal representation within a temporal ($t > 0$) space.

Before departing this section, I would stress that within our universe everything has a price to pay, a section of time Δt and an amount of energy ΔE and it is not free. Quantum qubit information pays no price, yet qubit information created a worldwide qubit conspiracy. From this it is hard to tell when this conspiracy would end, but I am confident to say that this fictious qubit supremacy would end soon since information transmission is supposed to be physically realizable.

4.9   DOUBLE-SLIT PARADOX

Instead of getting into the argument of simultaneous existence particles at double-slit using Young's experiment, this is a non-physical realizable paradigm from physically realizable standpoint. Particle-wave dynamics is a mathematical equivalent duality principle as described: a particle in motion is equivalent to wave dynamics or wave propagation is equivalent to particle dynamics. However, a particle is not equal to a wave and a wave is not equal to a particle. Particularly, as from the De Broglie-Bohm theory that I quote: particles have "precise locations" at all times … [10]. In contrast within a temporal ($t > 0$) subspace, a particle changes with time but "not" at a precise location since a future prediction is not deterministic. As we have shown earlier, a particle that existed within a temporal ($t > 0$) space is quite different, as assumed within a virtual non-physically realizable subspace. For example, a particle existed within our temporal ($t > 0$) universe, no matter how small it is, and it has to be temporal ($t > 0$). Since a temporal subspace is not empty, this is compacted with temporal substances, from which we see that a particle cannot be totally isolated. For example, a mass particle induces gravitational field, a charged particle induces an electric field, and others that cannot be ignored. Without the preexistent substances such as permittivity and permeability, wave dynamics has no way to exist. From this we see that particle-wave dynamics is a mathematical postulation that existed only within an empty timeless ($t = 0$) or time-independent virtual mathematical subspace, since the assumption of wave dynamics is not a time- and band-limited physically realizable wavelet.

Nevertheless, let me show a double-slit setup, as depicted in Figure 4.12(a), which is a commonly accepted paradigm that has been used for decades, but it is not a physically realizable paradigm. Yet a photonic particle can be shown simultaneously and instantaneously exists at the double slits, since within an empty space it has no time and no distance. And this is precisely the same subspace that Schrödinger's superposition principle derived from, but we have shown that the superposition principle can only exist within an empty timeless (t = 0) virtual subspace.

However, if the double-slit hypothesis is situated within a temporal (t > 0) subspace, as depicted in Figure 4.13(b), then it is very unlikely two particles will instantaneously and simultaneously exist at both slits because time is distance and distance is time within a temporal (t > 0) subspace. Since a wave is equivalent to a particle from a particle-wave dynamics standpoint, within our temporal (t > 0) universe any physical wave dynamics has to be time and band limited, otherwise it is a virtual wave dynamic. From this we see that it is very unlikely two wavelets (or particles) will simultaneously arrive at both slits at the same time.

Yet, a question remains to be asked: why does it work for a continuously emitting laser? It is apparent that a continuous light emitter has a longer time-limited duration. For example, if we assume that humans have a 300-year life expectancy, then it is a good chance that we may coexist with Einstein, Schrödinger, and may be Newton at some time, but may not at the same place. On the other hand, if our universe is a time-independent (i.e., timeless) space, then in principle we can time-travel back to visit them. What I have just given is that within our temporal (t > 0) universe, everything has a price; an amount of energy ΔE and a section of time Δt (i.e., ΔE, Δt) to pay. But this is the necessary cost, yet it is not sufficient. From this we see that the superposition principle is limited by a section of time Δt, although ΔE and Δt coexist.

Nevertheless, we can hypothetically show that instantaneously and simultaneously superposition phenomena do not hold by a postulated setup, as shown in Figure 4.14, which is a physically realizable paradigm since substance and temporal (t > 0) space are mutually inclusive.

However, if the different path length between d1 and d2 is beyond the coherence length D of the laser as given by:

$$D = d_2 - d_1 = c(\Delta t_2 - \Delta t_1) = c\ \Delta t' < D \tag{4.38}$$

(a)   (b)

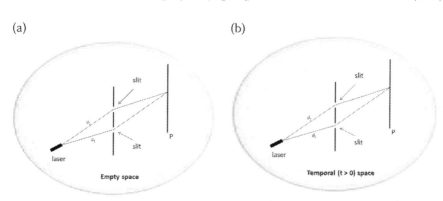

FIGURE 4.13 Shows a schematic diagram of a double-slit experiment. (a) Shows an empty space paradigm, (b) shows a physically realizable paradigm.

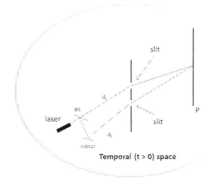

FIGURE 4.14 Shows a schematic diagram of a double-slit experiment using a band-limited coherent light source.

where d_s are the distances, Δt_s are the incremental times, and c is the velocity of light, then the interference pattern cannot be observed at the diffraction screen of P. This means that photonic-particles (i.e., photons) emitted from the laser are not simultaneously and instantaneously arriving at the double slit, as from the coherence theory of light (see Appendix F).

Let me further note that if one submerges any scientific model within a temporal (t > 0) subspace, then it is rather easy to find out the paradox, as observed within an empty subspace that does not exist. Notice that whenever a scientific model is submerged within a temporal (t > 0) subspace, the model becomes a part of the temporal (t > 0) space for analysis, from which many of the timeless (t = 0) paradoxes can be resolved rather easily; for instance, Schrödinger's cat and Einstein's theories. Nonetheless, this is an unintentional error that all scientists have committed for centuries. For instance, all the laws, principles, theories, and paradoxes were developed from the same empty timeless (t = 0) subspace. For this, most of the scientists believe that we can travel beyond and behind the pace of time, as Einstein's special theory has suggested. Similarly, we can simultaneously and instantaneously exploit photonic particles for computing and communication as Schrödinger's fundamental principle of superposition has promised.

For example, if one plunges two moving spaceships within an empty space, we cannot tell which one is moving with respect to the other. However, if we submerge the same scenario within a temporal (t > 0) subspace, inevitably we can figure out the relative position between them, since time is space and space is time within a temporal (t > 0) subspace, while within an empty subspace there is no time and no distance. And this is precisely why Einstein's special theory is relativistic-directional independent and as well his general theory of relativity is a deterministic principle that I shall discuss in Chapter 5. From this it is trivial for us to submerge a pair of entangled particles within a temporal (t > 0) subspace, then we would find out the instantaneous (i.e., $\Delta t = 0$) entanglement does not exist, since within our universe there is always a section of time Δt to pay aside an amount of energy ΔE, and these are not free.

Let me further stress that time speed is one of the most esoteric variables that exist with our universe that cannot be changed, but it is the section of time Δt that we spend that can be somewhat manipulated. From this we see that the section of Δt that we will spend can be squeezed as small as we wish yet we can never be able to squeeze it to zero (i.e., $t = 0$), even when we assume to have all the energy ΔE (i.e., $\Delta E \longrightarrow \infty$) willing to pay for. And this is the well-known causality constraint within our temporal (t > 0) universe that cannot be violated.

Furthermore, a question remains to be asked; if the width of Young's experiment is smaller than the wavelength of the illuminator, would you able to observe the diffraction pattern? If the answer is no, then we see that wave dynamics is equivalent to a particle in motion but not equal to a particle since a photonic particle has no size. From this we see that a particle in motion is equivalent to wave dynamics, but a wave is not a particle and a particle is not a wave.

4.10 REMARKS

I would remark that quantum scientists used amazing mathematical analyses with their fantastic computer simulations to provide very convincing results. But mathematical analyses and computer animations are virtual and fictitious, and many of their animations are "not" physically realizable; for example the "instantaneous and simultaneous" superimposing principle for quantum computing does "not" actually exist within our universe. One of the important aspects within our universe is that one cannot get something from nothing and there is always a price to pay: an amount of energy ΔE and a section of time Δt. The important thing is that they are not free!

Since any science that existed within our universe has time or is temporal (t > 0), in which we see that any scientific law, principle, theory, and paradox has to comply with a temporal (t > 0) condition within our universe, otherwise it will be unlikely the solution will be a physically realizable solution. Since science is mathematics but mathematics is not equal to science, I have shown that any analytic solution has to be temporal (t > 0); otherwise it cannot be implemented within our universe, which includes all the laws, principles, and theories.

Since Schrödinger's quantum mechanics is a legacy of Hamiltonian classical mechanics, I have shown that Schrödinger's mechanics is a timeless (t = 0) machine since Hamiltonian mechanics is timeless (t = 0). This includes Schrödinger's fundamental principle of superposition, which is "not" a physically realizable principle. Since Schrödinger's cat is one of the most controversial paradoxes in the modern history of science, I have shown that the paradox of Schrödinger's cat is "not "a physically realizable paradox, which should not have been postulated!

The most esoteric nature of our universe must be time, for which every fundamental law, principle, and theory is associated with a section of time Δt. We have shown that it is the section of Δt that we have used and cannot bring it back. And this is the section of Δt that a set of the most elegant laws and principles are associated. Of this I have shown that we can squeeze a section of time Δt close to zero (i.e., $\Delta t \to 0$) but it is "not" possible to reach zero (i.e., $\Delta t = 0$) even though we have all the energy ΔE to pay for it. In this we see that we can change the section of Δt, but we cannot change the speed of time.

Since quantum computing and communication rely on qubit information logic, but qubit information can only exist within a timeless (t = 0) subspace, for this I have shown that qubit information is as virtual and illusive as Schrödinger's cat. This is not a physically realizable qubit information that can be used for communication and as well for computing within our time-space.

Although the double-slit hypothesis is a well-accepted postulation for showing the superposition principle, unfortunately the postulation only holds within a non-physically realizable empty space paradigm, and it does not exist within our temporal (t > 0) universe. What I mean is that a double-slit postulation is another false hypothesis just as Schrödinger's cat, that has let us to believing superposition actually existed within our universe.

Overall, this chapter has shown that it is not how rigorous the mathematics is, it is the physically realizable paradigm that produces the physically realizable solution. From this we see that If one used a non-physically realizable paradigm, it is very "likely" one will get a non-physically realizable solution.

Finally, I would stress that the nature of temporal (t > 0) quantum mechanics is all about the physically realizable mechanics. From this we see that it is our mechanics that change with time, but it is our mechanics that stop the time. Nevertheless, I had shown that we can change a section of time Δt, but we cannot change the pace of time or even stop time as Schrödinger's wishes. Yet, it is the physically realizable science we embraced, but not the fancy mathematical solution we adored.

REFERENCES

1. A. Einstein, *Relativity, the Special and General Theory*, Crown Publishers, New York, 1961.
2. E. Schrödinger, "An Undulatory Theory of the Mechanics of Atoms and Molecules", *Physical Review*, 28 (6): 1049, 1926.
3. W. Heisenberg, "Über den anschaulichen Inhalt der quantentheoretischen Kinematik und Mechanik", *Zeitschrift Für Physik*, 43 (3–4): 172, 1927.
4. L. Boltzmann, "Über die Mechanische Bedeutung des Zweiten Hauptsatzes der Wärmetheorie", *Wiener Berichte*, 53: 195–220, 1866.
5. J. M. Knudsen, and P. Hjorth, *Elements of Newtonian Mechanics*, Springer Science & Business Media, 2012.
6. R. C. Tolman, *The Principles of Statistical Mechanics*, Dover Publication, London, 1938.
7. D. F. Lawden, *The Mathematical Principles of Quantum mechanics*, Methuen & Co Ltd., London, 1967.
8. F. T. S. Yu, "The Fate of Schrodinger's Cat", *Asian Journal of Physics*, 28 (1): 63–70, 2019.
9. N. Bohr, "On the Constitution of Atoms and Molecules", *Philosophical Magazine*, 26 (1): 1–23, 1913.1
10. E. MacKinnon, De Broglie's Thesis: A Critical Retrospective, *American Journal of Physics*, 44: 1047–1055, 1976.
11. F. T. S. Yu, "What is "Wrong" with Current Theoretical Physicists?", *Advances in Quantum Communication and Information*, Edited by F. Bulnes, V. N. >. Stavrou, O. Morozov and A. V. Bourdine, Chapter 9, pp. 123–143, IntechOpen, London, 2020.

12. C. H. Bennett, "Quantum Information and Computation", *Physics Today*, 48 (10): 24–30, 1995.
13. K. Życzkowski, P. Horodecki, M. Horodecki, and R. Horodecki, "Dynamics of Quantum Entanglement", *Physical Review A*, 65, 1–10, 2001.
14. Einstein Attacks Quantum Theory, *Scientist and Two Colleagues Find It Is Not 'Complete' Even Though 'Correct'*. The New York Times, April 28, 1935.
15. F. T. S. Yu, "Aspect of Particle and Wave Dynamics", *Origin of Temporal (t > 0) Universe: Correcting with Relativity, Entropy, Communication and Quantum Mechanics, Appendix*, CRC Press, New York, 145–147, 2019. New York.
16. C. E. Shannon and W. Weaver, *The Mathematical Theory of Communication*, University of Illinois Press, Urbana, IL, 1949.
17. F. T. S. Yu, "From Relativity to Discovery of Temporal (t > 0) Universe", *Origin of Temporal (t > 0) Universe: Correcting with Relativity, Entropy, Communication and Quantum Mechanics*, Chapter 1, CRC Press, New York, 1–26, 2019. New York.
18. E. Parzen, *Stochastic Processes*, Holden Day Inc., San Francisco, 1962.
19. F. T. S. Yu, *Optics and Information Theory*, Wiley -Interscience, New York, 1976. *Information Theory*, Wiley
20. D. Gabor, "Communication Theory and Physics", *Philosophical Magazine*, 41 (7): 1161, 1950.
21. F. T. S. Yu, "Information Transmission with Quantum Limited Subspace", *Asian Journal of Physics*, 27 (1): 1–12, 2018.
22. L. J. Cultrona, E. N. Leith, L. J. Porcello and W. E. Vivian, "On the application of Coherent Optical Processing Techniques to Synthetic-Aperture Radar", *Proceedings of the IEEE*, 54: 1026, 1966.
23. E. N. Leith and J. Upatniecks, "Reconstructed Wavefront and Communication Theory", *Journal of the Optical Society of America*, 52: 1123, 1962.
24. F. T. S. Yu, *Introduction to Diffraction, Information Processing and Holography*, Chapter 10, MIT Press, Cambridge, Mass, 91–98, 1973.
25. N. Wiener, *Cybernetics*, MIT Press, Cambridge, MA, 1948.
26. N. Wiener, *Extrapolation, Interpolation, and Smoothing of Stationary Time Series*, MIT Press, Cambridge, MA, 1949.
27. F. T. S. Yu, "Time: The Enigma of Space", *Asian Journal of Physics*, 26 (3): 143–158, 2017.

5 Nature of Relativity

As we all agreed that Einstein's space-time continuum is one of the most revolutionary ideas in modern science since he disclosed it over a century ago. But I have found his relativity theories were derived within an empty abstract subspace, for which his special and general theories are timeless (t = 0) or time independent. It is the reason I shall start this chapter with the nature of time as a dependent variable of space, since time and space coexist [1,2]. As in contrast with most of scientists that have believed time is an independent dimension with space, firstly, Einstein's special theory of relativity shows no sign of direction, and I will show that (as momentarily accepted his theories) it is in fact a relativistic-directional dependent theory instead of a directional independent principle. Since the big bang theory should have been started within a preexistence temporal (t > 0) space, I will show an avoidable induced gravitational field due to mass that should be added with the energy equation to ignite the explosion. Since the empty subspace paradigm has been used for centuries, I have found that scores of fictitious theories and virtual principles have merged; for example, time-image symmetric principle, anti-matter, and negative energy are virtual and illogical and cannot actually exist within our temporal (t > 0) universe. I will also show that Einstein's special theory and his general theory of relativity are not physically realizable theories. Of this we see that it is not how rigorous mathematics is; it is the physically realizable science we embrace. Nevertheless, we learned that it is the empty subspace that we had used for centuries that had undermined the physical reality of our science.

5.1 NATURE OF TEMPORAL (T > 0) SPACE

As we accepted subspace and time coexists within our temporal (t > 0) universe, time has to be real, and it cannot be virtual since we are physically real. And every physical existence within our universe is real. The reason some scientists believed time is virtual or an illusion is that it has no mass, no weight, no coordinate, no origin, and it cannot be detected or even be seen. Yet time is an everlasting existing real variable within our known universe. Without time there would be no physical matter, no physical space, and no life. The fact is that every physical matter coexists with time, which includes our universe. And the reality of time is that every physical substance changes with time for which we see that it is our universe that changes with time, and it is not our universe that changes the time. In other words, we walk on the street; it is not the street that walks us. Therefore, when one is dealing with science, time is one of the most enigmatic variables that is ever present and cannot be simply ignored. Strictly speaking, all the laws of science as well every physical substance cannot exist without the existence with time. For this we see that time cannot be an independent dimension or an illusion. In other words, if

DOI: 10.1201/9781003271505-5

time is an illusion, then time will be independent from physical reality or from our universe. And this is precisely how many scientists have treated time, as an independent variable, such as Minkowski's space [3], for which space can curve time (e.g., time-space can be changed by gravity). However, if matter can curve time-space, then we can change the speed of time. But we see that it is our universe that exists with time, and it is not our universe that changes the time. In other words, time is a forward-dependent variable with every substance that includes our universe. For this we cannot change the pace of time or even stop time.

Since time is a forward moving dependent variable at a constant pace, for scientific presentation we have used symbolic substitution to represent time; otherwise it would be very difficult to facilitate and to understand the nature of time. For convenience, we divided time into past (i.e., t < 0), present (i.e., t = 0), and future (i.e., t > 0) domains to represent time, as summarized in Figure 5.1.

In this it shows that our universe changes with time; for example, the present moment at t = 0 moves immediately forward to become the next present moment t + Δt. In other words, the present moment t = 0 becomes the moment of the past. Once the present moment (t = 0) moves a section of Δt ≈ 0, no matter how small it is, it is impossible to return back, since our universe changes with time. From this we also see that it is impossible to move the current moment (t = 0), no matter how small Δt is, ahead or behind the pace of time. Nevertheless, this diagram epitomizes our temporal (t > 0) universe changes with time, since our universe is a stochastic dynamic temporal (t > 0) space [1,2]. From this we see that it is impossible to travel backward or ahead of the pace of time.

Since the past-time domain (i.e., t < 0) represents the moment of certainty events, they were the past memories (i.e., information) but without physical substance in it. This is similar to viewing a backward video clip. In other words, if we move time backward (t < 0), we see those past consequences (i.e., past universes) that change

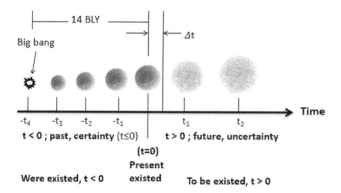

FIGURE 5.1   Shows that our temporal (t > 0) universe changes naturally with time. From this it shows that the age of our universe is about 14 billion light years old. The past-time domain (t < 0) represents a set of certain virtual events, the future-time domain (t > 0) represents a couple of physically realizable uncertainty events. The present absolute certainty moment (t = 0) will be instantaneously moved forward to become the next new present moment t = 0 + Δt, where Δt ≈ 0.

precisely with time (e.g., $t = -t_n$), as a backward movie clip. From this we see that it is time changing the videos (e.g., past events) but it is not the video changing our time.

In view of Einstein's general theory of relativity: matter (i.e., time-space) curves (or changes) time-space. And this is precisely the section of time (i.e., $t < 0$) that the general theory developed from, since the general theory treated time as an independent variable (i.e., timeless $t = 0$). For this general theory, it is a deterministic theory similar to all the deterministic classical science (e.g., $f = ma$, $E = mc^2$ and others).

Nevertheless, within our temporal ($t > 0$) universe, time is a dependent variable with respect to the subspace since space and time coexist. In this we see that future events are non-deterministic consequences, but with some degree of certainty. And this is precisely the positive time [i.e., temporal ($t > 0$)] domain where general theory may not apply, since any future space is not deterministic or change naturally with time. Nevertheless, the implication of temporal ($t > 0$) is that physically realizable events exist if and only if within the positive time domain, by which the instantaneous $t = 0$ can only be an approach but will never be able to attain (i.e., $t \longrightarrow 0$), even we have all the energy (i.e., ΔE) to spend.

Since science exists (or changes) with time, for which we see that time is one of the most intriguing variables within our universe. To understand the nature of time, firstly, we have to understand the creation of our temporal ($t > 0$) universe, as illustrated in Figure 5.2.

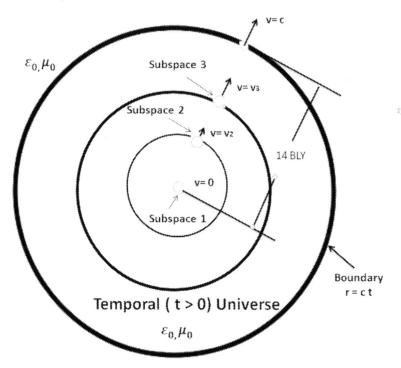

FIGURE 5.2   Shows a dynamic temporal universe: v represents the velocity of the subspace, c is the speed of light, (ε_0, μ_0) are the permittivity and permeability of the deep space, BLY represents billion of light years.

This simple diagram epitomizes the dynamic behavior of a temporal (t > 0) stochastic universe in which its boundary expands rapidly at the speed of light. Our universe was created by a well-accepted big bang theory that happened about 14 billion light years ago based on the Hubble space telescope observation [4]. Within the expanding universe, it dramatizes a set of its subspaces moving outwardly toward the boundary at different velocities, and the speed of the subspaces is linearly proportional to the speed of light as the boundary of our universe expands at the speed of light. Since these subspaces represent the past light years of our universal dynamic history, this can be shown in a temporal (t > 0) representation of the past, as described in Figure 5.3.

This diagram shows the past 14 billion light years (BLY) of our universe is frozen at the present moment t = 0. This figure represents a time reversal in a temporal (t > 0) domain, which shows that our universe is continuingly expanding with time. In other words, without the expansion of our universe, it would be very difficult to determine the velocity of the subspaces within our universe if our universe is a bounded static universe, instead of a dynamic expanding universe.

Since the outward velocity of the subspace is proportional to the speed of light within the section of 14 BLY, the respective velocity of the subspace can be extrapolated. Using this limited amount of information from the Doppler shift, it is sufficient to determent each subspace's location. Since the Doppler shift provides us a mean to determine the location of a moving subspace, it does not give us the detail of the subspace, although past consequences are certainty events. From this we note that the Doppler shift detection does not have the same scenario as traveling backward in time as the special theory of relativity has promised. Yet I shall show subsequently that Einstein's special theory is a not a physically realizable principle within our temporal (t > 0) universe since we cannot change time, although we change with time. In other words, we had went with the time but the time does not go with us.

Since time coexists with subspace, any subspace within our temporal (t > 0) universe cannot be empty and the speed of time is the same everywhere within our universe. But with reference to special theory, it means that the speed of time within

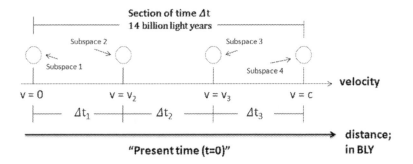

FIGURE 5.3 Shows a temporal representation of the past. It shows a set of subspaces in preceding Figure 5.2 as a function of velocity presented at current time t = 0, where their respective distances can precisely identify Δt_n in terms of BLY (i.e., billions of light years).

a moving subspace is relative with respect to relative motion of the subspaces, from which we see that there is a contradiction within our universe since every subspace has the same time regardless of their velocities. Nevertheless, our temporal (t > 0) universe was created by Einstein's energy equation although physically correct but not consistent since $E = Mc^2$ was derived from his special theory. As we accepted Einstein's energy equation, our universe was created by a big bang theory using his energy equation as given by [1,2]:

$$\frac{\partial E(t)}{\partial t} \approx -c^2 \frac{\partial M(t)}{\partial t} = [\nabla \cdot S(t)], \quad t > 0 \tag{5.1}$$

where ∇ represents a divergent operator, $S(t)$ is a thermo-nuclei energy vector, and (t > 0) denotes the equation is subjected to the temporal (t > 0) constraint. From this presentation we see that every subspace (or substance) is temporal (t > 0), and every subspace is created by an amount of energy ΔE and a section of time Δt (i.e., $\Delta E, \Delta t$). From this we see that in principle the section of Δt can approach zero (i.e., $\Delta t \longrightarrow 0$), but never be able to reach the instantaneous instance (i.e., t = 0). This section of Δt will be in the part of the nature of time since a section of time Δt is profoundly connected to every physical aspect of our universe.

In order to epitomize the temporal (t > 0) nature of our universe, I came up with a composite diagram, as depicted in Figure 5.4, which shows that the nature of our temporal (t > 0) universe was started from a big bang creation, although time has

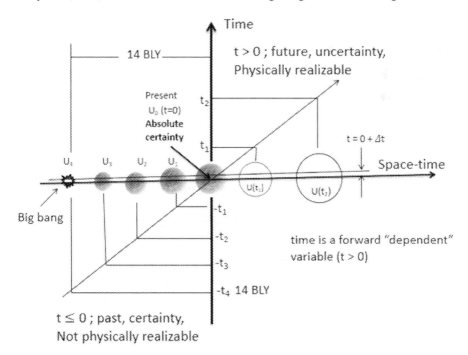

FIGURE 5.4   Shows a composited temporal (t > 0) time-space diagram to epitomize the nature of our temporal universe. BLY is billions of light years.

existed well before the creation. Since the past certainty's consequences (i.e., memory spaces) happened at a specified time within the negative time domain (i.e., t < 0), we see that every specific past time event has been determined with respect to a precise past certainty subspace. From this we see that time can be treated as an independent variable with respect to the past certainty consequences within the pass-time domain (t < 0) as from a mathematical standpoint.

Nevertheless, from a physical reality standpoint, time is longer existed within the negative time (t < 0) domain. And this is precisely how time has been treated as an independent variable from a mathematical standpoint to predict the distant future with some degree of certainty, since physical substance or subspace changes naturally with time within our temporal (t > 0) space. This is precisely what classical laws and principles have done to science, using past deterministic certainties to predict the future. This is exactly the reason why deterministic prediction (i.e., independent from time) comes from a deterministic analysis. But within our temporal (t > 0) universe, strictly speaking, everything is unpredictable. In other words, the farther away from the absolute instant moment (i.e., t = 0) of certainty, the more unpredictable. From this we see that all physically realizable solutions strictly speaking should be temporal (t > 0) or a predictable solution with degree of uncertainty. From this we see that Einstein's general theory of relativity is a deterministic principle that is not a physically realizable theory.

Although using past certainties to predict a future outcome is a reasonable method that had been used for centuries, but it is physically wrong if we treated time as an independent variable within our temporal (t > 0) universe. And this is the reason scores of irrational and fictitious solutions emerged, that has already dominated the worldwide scientific community. This includes Schrödinger's fundamental principle of superposition [5], Einstein's special and general relativity theories [3], and many others, since they were all based on past certainties to predict a deterministic future, which is not a temporal (t > 0) solution or changes with time (i.e., non-deterministic).

Yet, the section of time Δt shown in Figure 5.4 represents an incremental moment after the instant t = 0, which moves to a new t = 0 + Δt, where in principle Δt can be squeezed as small as we wish (i.e., $\Delta t \longrightarrow 0$). But we will never be able to squeeze it to zero (i.e., $\Delta t = 0$) even if we assumed we have all the energy ΔE to pay. In fact, this is the section of time that cannot be delayed or moved ahead the pace of time (i.e., t < 0 + Δt or t > 0 + Δt). From this we see the aspect for time traveling either ahead or behind the pace of time is not conceivable, since we coexist with time.

5.2 ASYMMETRIC PRINCIPLE

The basic distinction between science and mathematics is that one is physical realizability and the other is virtual reality. Since science is mathematics, then its analytical solution has to be physically realizable. From this I have found that the mirror symmetric principle is not a physically realizable principle, since the principle is based on a mirror image of time (i.e., t < 0) or negative time domain. For example, if we allow an anti-matter to exist within our temporal (t > 0) universe, then we should have an anti-anti-mater that existed within the virtual negative time (t < 0) domain and so on and so on. Nevertheless, a negative time (t < 0) domain cannot exist within our temporal (t > 0) by virtue of the temporal (t > 0) exclusive principle (see Appendix A, E).

Our universe changes with time, but it is not a mirror image of time as some scientists claimed [6]. The reason is that the past-time domain (i.e., t < 0) has no physical substance and has no time. In this we see that the present moment (t = 0) of our universe is the if and only if absolutely physical certainty. Yet the present moment of t = 0 will be instantly moved away to become the next moment of a new instant moment of t = 0 + Δt. From this we see that it is impossible for us to go back or ahead of the pace of time.

Yet time and universe coexist, from which we knew that time is not an independent variable as most of the scientists have assumed. As I have noted earlier, the reason of treating time as an independent variable was that our analyses had mostly been derived on a piece of paper that had been inadvertently treating its background as an absolute empty timeless (t = 0) space for centuries. Since substance and emptiness are mutually exclusive (i.e., temporal exclusive principle), I had found practically all the scientific laws, principles, and theories strictly speaking are not physically realizable. For example, if we applied our laws or principles directly within our temporal (t > 0) universe, I had found practically all of them are not temporal (t > 0) equations, but virtual and fictitious as mathematics does.

In view, the nature of time of any subspace within our universe is temporal (t > 0), from which we see that any temporal (t > 0) subspace cannot have time speed either faster or slower than the pace of time; otherwise the subspace cannot exist within our universe. For example, any gravitational field induced by a temporal mass m(t) within our universe has the same time-speed as our universe. Since our universe is compacted with temporal (t > 0) substances that include all the permeability and permittivity [i.e., $\mu_0(t)$ $\varepsilon_0(t)$] the non-particle form substances, from this we see that every mass m(t) is capable of creating a temporal gravitational field F(t) induced by mass m(t) within any temporal (t > 0) space.

Since a symmetric principle (mirror image) comes from a virtual Newtonian subspace, where time has been treated as an independent variable [7], such that the negative time domain is included. Yet temporal (t > 0) and non-temporal are mutually exclusive; any non-temporal space cannot exist within our universe. And this is the reason that mirror-image symmetric principle of physics cannot exist within our temporal (t > 0) universe. This is in contrast with some scientists who believed that a timeless (t = 0) subspace is an accessible subspace within our universe. From this we see that anti-matter is as virtual as mathematics is; for instance, the symmetric principle of science can be summarized as follows [8] (see Appendix F):

Physical Reality versus Image Virtual Reality

Positive Time (t > 0) versus Negative Time (t < 0)

Positive Energy versus Negative Energy

Matter versus Anti-Matter

And others

Since all mirror images of the symmetric principle located in the negative time (t < 0) domain are virtual, from this we see that they are naturally excluded within our temporal (t > 0) domain; for example, negative energy, anti-matter, and others. Nevertheless, if there is an anti-matter or a negative energy substance within our temporal (t > 0) space, then would we have the anti-anti-matter or negative-negative energy matter within the negative time (t < 0) space? From this we see that it would be more physically reliable to search any new particle that is temporal (t > 0), since a non-temporal or timeless (t = 0) particle cannot exist within our universe. In other words, it is kind of wasting our precious effort to search a non-existent anti-particle that can only exist within the virtual (negative) time domain (t < 0). Nevertheless, our universe is an energy conservation subspace that expands at speed of light. It is however not a zero-summed energy subspace that allows anti-matter to exist. From which we see that, four-dimensional space-time continuum is a zero-summed energy space. Precisely, this is the reason that anti-matter was theoretically hypothesized by Dirac [15] Yet, his hypothesis exists only within a zero-summed empty subspace, which cannot be existed within our temporal (t > 0) universe where time is a dependent variable.

Since every temporal (t > 0) subspace is compacted with substances that also includes all the non-particle form substances, we see that a temporal (t > 0) universe exhibits an asymmetric principle, which is the temporal (t > 0) physical reality versus the timeless (t = 0) virtual reality, as given by:

Have versus Have Not

Temporal (t > 0) reality versus Timeless (t = 0) virtual reality

Energy (E > 0) versus No Energy

Matter versus No Matter

And others

From this we see that the asymmetric principle of science is based on a dependent forwarded time variable principle within our temporal (t > 0) universe, as in contrast with the symmetric principle where time is treated as an independent dimension. Therefore, if one is searching for any physical particle within our universe, it is more viable to look into the asymmetric principle instead of the symmetric principle since every physical particle within our universe is a temporal (t > 0) particle. Otherwise, it is like searching for a timeless (t = 0) angle particle that does not exist within our temporal (t > 0) universe, but only existed within a virtual timeless (t = 0) or time-dependent mathematical subspace.

5.3 QUEST OF RELATIVISTIC TIME

Einstein's special relativity theory was derived using Minkowski's four-dimensional space-time continuum with Lorentz transformation [3]. In other words, the special theory

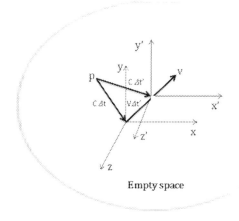

FIGURE 5.5 Shows a Minkowski's four-dimensional space-time continuum with Lorentz transformation. From this, Einstein's special theory of relativity was developed.

was derived on a Euclidian-based coordinate system translating at a constant velocity v within an empty subspace (i.e., a piece of scratched paper), as depicted in Figure 5.5.

Although the empty space paradigm is not a physically realizable model, yet Einstein's special theory is one of the most revolutionary principle that was developed at the dawn of the modern physics era in the past century as given by:

$$\Delta t' = \Delta t / [1 - (v/c)^2]^{1/2} \qquad (5.3)$$

where v is the velocity of a coordinate system and c is the speed of light. Nevertheless, the theory shows no sign of relativistic direction, although the implication is relative-directional. This is similar to the kinetic energy equation, it has no sign of direction, but the equation implies that its energy vector is at the same direction of the velocity vector. From this we see that it is wrong to treat a special theory as a relativistic-directional independent.

Nevertheless, if a special theory is developed within a temporal (t > 0) subspace, as depicted in Figure 5.6, an induced gravitational field by a moving mass M cannot be simply ignored. From this we see that a huge gravitational field induced by a moving mass (or particle in motion) travels at the same velocity with the mass. Since a moving mass is heavier than a rest mass, according to Einstein's special theory, but its induced gravitational field has never been a part included in the big bang explosion. As I shall show subsequently that the induced gravitational field contributed a significant factor for big bang explosion, as well for any giant star explosion that includes a possibly black hole annihilation. From this we see that an induced gravitational field plays an important role as we are dealing with high-speed particle acceleration, as well as for a very large thermo-nuclei mass annihilation such as that applied to gravitational wave detection [9].

Similarly, Einstein's special theory can be developed within a temporal (t > 0) subspace, as depicted in Figure 5.6. By ignoring the induced gravitational field of

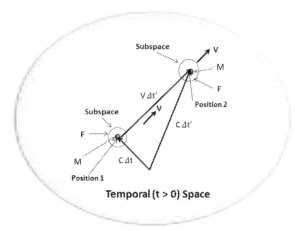

FIGURE 5.6   Shows Einstein's relativity equation can be derived within a temporal (t > 0) subspace. F represents the induced gravitational field of mass m.

F momentarily, with reference to the Pythagorean theorem, as illustrated in the diagram, we have the following special theory relationship:

$$\Delta t' = \Delta t/[1-(v/c)^2]^{1/2}, \quad t > 0 \text{ and } v \rightarrow \tag{5.4}$$

where t > 0 denotes that the equation is subjected to temporal (t > 0) constrain and v $\longrightarrow$ represents the velocity vector. This shows that the special theory of relativity is a relativistic directional principle with respect to velocity vector v. Notice that I had intentionally ignored the relativistic with respect to the velocity vector of c.

Although the temporal (t > 0) relativistic equation is identical to Einstein's special theory, the implication shows it is relativistic-directional dependent. For example, if a traveling subspace of velocity v passes position 1 at time $t = t_1$ and at the same instant a light beam starts to illuminate the static position 1, then the whole derivation process is started at $t = t_1$, which is the actual time of the temporal (t > 0) subspace. From this we see that it will take a section of time $\Delta t'$ for the moving subspace to reach position 2, that is the time of the temporal (t > 0) subspace at $t_2 = t_1 + \Delta t'$, which is exactly equal to the time of subspace that has changed from t_1 to t_2. From this we see that there is a section of relativistic time-gain $\Delta t''$ (i.e., $\Delta t'' = \Delta t' - \Delta t$) or time-lost, which is depending on how we interpret the theory. Nevertheless, the legitimacy of Einstein's relativity theories is a topic that I shall discuss in a subsequent section.

Firstly, if we interpret the section of $\Delta t''$ as a relativistic time-gain with respect to static position 1, then we should interpret as a relative time-lost with respect to the position 2 since within our temporal (t > 0) universe one cannot get something from noting there is always a price to pay, for example a section time and an amount of energy (i.e., $\Delta t, \Delta E$). In other words, any relativistic section time-gain $\Delta t''$ cannot be retained or to keep unless the moving subspace can instantly return back to the original static position 1. This is due to the fact that within our temporal (t > 0) universe we can neither change nor stop time. The relative section of time-gain $\Delta t''$ within the moving subspace will eventually lose it as I shall show in the following.

On the other hand, if the scenario of Einstein's special theory is derived within an empty timeless ($t = 0$) subspace instead, the moving space can return back instantly (i.e., $t = 0$) to the original static subspace at position 1, since within empty space it has no time to lose. From this we see that a section of time gains $\Delta t''$ is added to static subspace at position 1 [or any static subspace by virtue of Eq. (5.3)]. Because the moving subspace can instantly return back to position 1 within an empty timeless ($t = 0$) space, then we see that the traveler would have a section of time gain $\Delta t''$ ahead of any static subspace.

But if the traveling scenario is assumed within a temporal ($t > 0$) subspace, the moving subspace (i.e., the traveler) supposes a section of relative time-gain $\Delta t''$ (i.e., $\Delta t'' = \Delta t' - \Delta t$) when the traveler reaches static position 2, as based on Einstein's special theory. However, if the traveler returns back to static position 1 at the same velocity of v he will eventually lose all the section of time gain $\Delta t''$ that he had gained when he reaches static position 2, since relativistic situation is opposite to the approaching as departing a static position within a temporal ($t > 0$) space.

As in contrast within an empty timeless ($t = 0$) space scenario, the relativistic time-space is independent with time, since within an empty space it has no time and no direction. But within our temporal ($t > 0$) universe it has time and has direction. From this we see that within our universe one cannot get something from nothing. There is always a price to pay an amount of energy ΔE and a section of time Δt (i.e., ΔE, Δt). This is the necessary cost constraint but is not sufficient [1,2]. From this we show Einstein's special theory is a relativistic directional principle, as in contrast with most scientists who assume it is a relativistic directional independent theory.

Let us get back to the section of relativistic time-gain $\Delta t''$ within a traveling subspace scenario, although relativistic time-gain has been a well-accepted theory as opposed to relativistic time-lost. But, as from the physical dynamics of our temporal ($t > 0$) universe, as depicted in Figure 5.7, we see relativistic with respect to the static subspace at the center of our universe should interpret as relativistic time-lost since the age of our observable universe is about 14 BLY ago. From this the outward moving subspaces represent the past-time domain (i.e., $t < 0$) of our universe which is about 14 billion light years, as estimated by the Hubble space telescope observation [4]. To be precise, this present moment of 14 BLY moves forward instantly by an incremental amount of Δt to be the next present moment (i.e., $t = 0 + \Delta t$), where Δt is an incrementally amount that approaches to zero (i.e., $\Delta t \rightarrow 0$).

Since all the moving subspaces are moving away from the center static subspace, a duration of Δt time has expended within the center static subspace is relatively to a section $\Delta t'$ within any moving away subspace (i.e., $\Delta t' > \Delta t$). In view of conservation of energy, the following uncertainty principle can be written as:

$$\Delta E' \ \Delta t' = \Delta E \ \Delta t = h \tag{5.5}$$

where we see that $\Delta E' < \Delta E$ since $\Delta t' > \Delta t$ and h is Planck's constant. Equivalently this means that lower $\Delta E'$ has a longer wavelength and narrower bandwidth as given by:

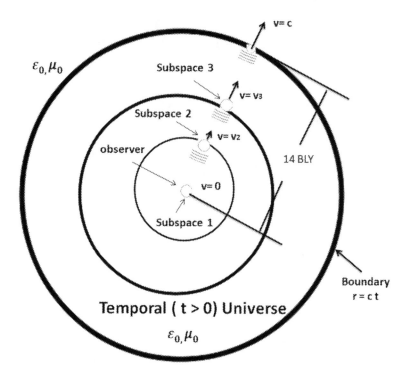

FIGURE 5.7 Shows that our current temporal (t > 0) universe is about 14 billion light years (BLY) of age. v represents the velocity of the subspace, c is velocity of light, (ε_0, μ_0) are the permittivity and permeability of the deep space, and c is the speed of light.

$$\lambda' > \lambda \qquad\qquad (5.6)$$

$$\Delta v' < \Delta v \qquad\qquad (5.7)$$

Since the section of relativistic time $\Delta t'$ is larger than the section of Δt within the statics subspace at the center of the universe (i.e., $\Delta t' > \Delta t$), the section of $\Delta t''$ represents a relatively time-lost (i.e., $\Delta t' - \Delta t'' = \Delta t$) with respect to the center static subspace. Or equivalently the relativistic time of the moving subspace with respect to the static center subspace that moves faster than the pace of time instead of slower as normally assumed. However, the time speed within any of the moving subspace remains the same as well within the static subspace. In other words, everywhere within our universe has the same time speed within the entire universe, although relativistic time between subspaces may be different, as a special theory described. But we note there is a discrepancy of Einstein's special theory of relativity with the physical reality of our temporal (t > 0) universe, since any moving subspace has a section of time gain $\Delta t''$ ahead of the center static subspace that has the same time as the entire universe of 14 BLY. In this we see that it is impossible for a moving subspace that has a section of time gain $\Delta t''$ ahead of the subspace itself, since the time speed of the subspace is the same within our universe. Once again, I have shown that it is impossible to travel ahead of the pace of time as in contrast with the special theory says it could.

Nevertheless, a subspace near the edge of the expanding universe has a relativistic time-lost with respect to the static center subspace that is equal to about 14 billion light years old (i.e., $t_1 \approx 14$ BLY) as can be seen in Figure 5.7. This is equivalent to a section of time Δt that our universe has changed with time at the speed of light. Yet the speed of time within the subspaces at the expanding boundary remains constant as well within the center static space. From this we see that the section of relativistic time-lost is a section of time that has been used, as observed from the static subspace toward the edge of the boundary. And this is the close that I can explain: the nature of time within our temporal ($t > 0$) universe.

On the other hand, if we treated the relativistic time lost as a gain with respect to the static center subspace, then the relativistic time within the moving subspace with respect to the standstill subspace moves slower as commonly assumed. Since it is a section of relativistic time-gain Δt of a moving subspace ahead of the static center subspace, where the actual time within the static subspace and the whole temporal universe has the same time. For this one can imagine if the section of relativistic time-gain of a moving subspace can instantly fly back to the standstill subspace at the center of the universe, then we would see that the traveling subspace will have a section of time-gain $\Delta t''$ added to the time within the static subspace. This means that a section of relativistic time-gain of $\Delta t''$ will be ahead of the actual time of the universe, then it is possible traveling to the future in time. But the instant fly back only can happen within an empty timeless ($t = 0$) space, but not within our temporal ($t > 0$) universe.

Besides, it is not physically possible to preserve the section of relativistic time-gain with respect to the center static subspace; there is a temporal ($t > 0$) issue of the relativistic time-gain if we carefully investigate the nature of temporal ($t > 0$) universe; as depicted in Figure 5.8, we see that a section of relativistic time-gain $\Delta t''$ is added to the current time of our universe $t + \Delta t''$ (i.e., 14 BLY + $\Delta t''$), which is a section of time $\Delta t''$ that our universe has not arrived at yet (i.e., $t = 14$ BLT). From this we see that the section of relativistic time-gain $\Delta t''$ represents a section of time gain ahead of our current time of our universe. Thus, we see that no matter how small $\Delta t''$ is, it is impossible for a traveler to move ahead of the current time even just for a very small fraction of time-gain $\Delta t''$ (i.e., $\Delta t'' \approx 0$).

But, if we treat a section of time Δt that has been used as a part of a price (i.e., ΔE, Δt) that our universe has gone through for its 14 billion light years creation after the big bang explosion, then we can view the past time 14 BLY of our temporal ($t > 0$) universe, as depicted in Figure 5.9. By referring to Einstein's special theory we see that time dilation for each outward moving subspace can be extrapolated, for which each corresponding section of time lost ($\Delta t'' = \Delta t' - \Delta t$) within the subspace can be extrapolated. For instance, a moving outward subspace near the edge of our universe can be calculated with the help of Doppler shift observation. For this a section of relativistic lost time $\Delta t''$ within the moving subspace can be determined. If we further assume $\Delta t''$ is treated as a section of relativistic lost time $\Delta t''$, then the actual time is 14 BLY $- \Delta t''$, which is slightly behind the pace of time, as can be seen in figure. But it belongs to a past-time universe of 14 BLY $- \Delta t''$, which no longer exists within our temporal ($t > 0$) universe.

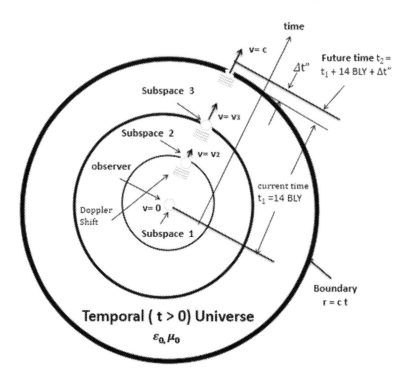

FIGURE 5.8 Shows a relativistic time-gain $\Delta t''$ temporal (t > 0) universe: our universe is about 14 billion light years (BLY) of current age plus a section of relativistic time-gain $\Delta t''$. v represents the velocity of the subspace, c is velocity of light, and (ε_0, μ_0) are the permittivity and permeability of the deep space.

Since every moving subspace has a section of time-lost $\Delta t''$ with respect to the center static subspace, if we assume a traveler within this moving space is able to fly instantly without any additional time expenditure into the static center subspace, we anticipate that the traveler would have a section of time-lost $\Delta t''$ with respect to the static subspace at the center of the universe. In other words, the traveler's time would be 14 BLY $- \Delta t''$, which is a short journey of $\Delta t''$ of the past with respect to current 14 BLY.

But we see that there is a section of relativistic time-lost $\Delta t''$ in every subspace within our universe since every subspace has the same time as our universe (i.e., 14 BLY). In other words, the whole universe has to move backward in time by a section of $\Delta t''$ in order for the traveler to preserve his time lost when he arrived at the center static subspace. In this we see that it is a section of relativistic time-lost $\Delta t''$ with respect to any subspace within our universe since all the subspaces have the same time as the universe.

Nevertheless, if our universe is situated within an empty space, then we see that the section of relativistic time-lost $\Delta t''$ can take it back instantly to any subspace that is equal to the current time (i.e., 14 BLY) minus a section of $\Delta t''$, which was the past time (i.e., 14 BLY $- \Delta t''$) of our universe. This means that going back to the

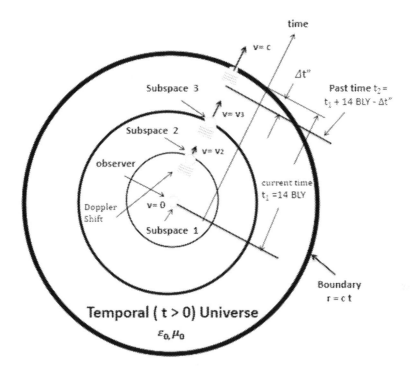

FIGURE 5.9 Shows a relativistic time-lost temporal (t > 0) universe, which is about 14 billion light years (BLY) current age minus a small section of relativistic time-lost $\Delta t''$. v represents the velocity of the subspace, c is velocity of light, and (ε_0, μ_0) are the permittivity and permeability of the deep space.

past (14 BLY – $\Delta t''$), as depicted in Figure 5.9. But our universe has left $\Delta t''$ to 14 BLY of the current moment. From this we see that the section of relativistic time-lost toward the past is impossible within our temporal (t > 0) universe. This example also shows us that within an empty space it has no price to pay, since it has no time and no space. But we are not living within a timeless (t = 0) virtual space.

Einstein's special theory of relativity shows no sign of relativistic directional, but we have shown it does in the preceding. For example, we assume a photonic traveler who has the advantage of traveling at speed of light starts an around the world voyage from Singapore, which is near the equator. By moving eastwardly around the world several times, the traveler is hoping that he will come back to Singapore relatively younger. Since the speed of light can travel around the world about seven times per second, from whom the traveler is hoping that Einstein's special theory is correct in order to prove the time traveling is the truth. Since it is a complicated geometrical problem, for simplicity we have flattened the globe several times to simply the hypothesis, as depicted in Figure 5.10. From this we see that the traveler has repeatedly circled the world eastward several times hoping to experimentally proven that Einstein is correct. Yet the photonic traveler comes home disappointed with no time gain since the special theory turns out to be a relativistic directional dependent theory instead of a directional independent, as a special theory represented in Eq. (5.3).

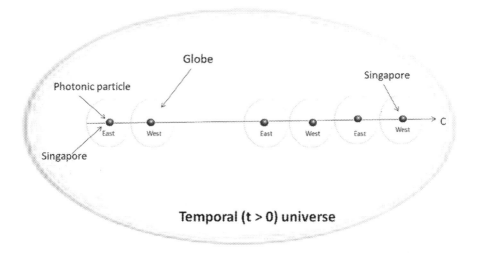

FIGURE 5.10   Shows a paradox of a photonic traveler's voyage at the speed of light.

Nevertheless, the photonic traveler's episode can also be understood in view of the nature of the temporal (t > 0) universe, as I have given earlier (i.e., Figure 5.4). Once again, we see that it is not possible for any substance or subspace within our universe (i.e., no matter how small it is or how fast it is) to move ahead or behind the pace of time with respect to the present moment t = 0. In this we also see that no matter how small the particle is, it cannot stop the time at t = 0 or instantaneously response at time t = 0, but it is possible to respond within a very small section of time Δt right after t = 0 (i.e., Δt → 0). From this we see that a particle behaves virtually the same within a micro-space as well in a macro-space, as in contrast with a generally accepted idea that a particle behaves differently within a micro-space.

5.4   NON-PHYSICALLY REALIZABLE SPECIAL THEORY

Nevertheless, either the section of relativistic time-gain or -loss (i.e., Δt″), the special relativity theory is still not consistent with the physically realizable principle within our temporal (t > 0) universe. For example, when a moving subspace reaches position 2, as referred to in Figure 5.6, the actual time of the static subspaces at position 1 or 2 is $t_1 + Δt'$, which t_1 is the static subspace's time when the moving subspace departed position 1. From this there is a missing section of time Δt″ (i.e., Δt″ = Δt′ − Δt) that has been ignored within the static subspaces I and 2, since the pace of time is the same within every subspace whether it is in motion or not. Thus, we see that a special theory has contradicted itself since a moving subspace would have a section of time-gain Δt″ (or lost) ahead (or behind) the moving subspace itself, where every subspace within our temporal (t > 0) universe has the same time. From this I cannot see there is a section of relative time-gain (or loss) from the special theory of relativity. It is therefore that we see that Einstein's special theory fails to exist within our temporal (t > 0) universe.

Since we had shown Einstein's special theory is not a relativistic principle in time, then Einstein's special theory is relativistic to what? In view of the temporal $(t > 0)$ paradigm of Figure 5.6, we see that Einstein's special theory is a relativistic distance instead of relativistic to time as given by:

$$d_r = (c-v)\Delta t' \tag{5.8}$$

where d_r is a relativistic distance that had been expended between an assumed light beam started at position S and a particle simultaneously started at position 1 moving at a constant velocity of v, from which we see that the moving particle and the light beam reached simultaneously at position 2. By this the light beam had traveled an extra distance of $d = (c - v)\Delta t'$ more (or relative) than the particle had had traveled. Thus, we see that Einstein's special theory is relativistic to distance within our temporal $(t > 0)$ subspace, instead of relative to time since we cannot change time. This means that a particle and the light beam simultaneously arrived at position 2 at the same time of the temporal $(t > 0)$ subspace, which is the time of every subspace within our universe that includes position 1, 2, and S. Once again, we have shown that there is no time-gain or time-loss of a traveling particle regardless of its speed.

Nevertheless, one of the misleading interpretations of Einstein's special theory must be directional independent. Although the special theory shows a velocity component in it (i.e., v^2), we had treated a special theory as a directionally in-dependent principle. Yet, within the equation it has a quadratic form of velocity [i.e., $(v/c)^2$] which had misled us over a century since Einstein disclosed the theory in 1905 [3]. Yet, when the velocity of the moving particle approaches the speed of light (i.e., $v \rightarrow c$), we have a relative distance $d_r \rightarrow 0$. This is by no means that time is running behind or ahead of the pace of time. But it is the speed of light that travels with time, and it is not the speed of light that changes the pace of time.

Similarly, relativistic distance can also be shown in terms of relative velocity of two moving particles. For example, two particles are moving in the same direction but at different speeds (e.g., v_1 and v_2). Aside, Einstein's special theory is not a physically realizable theory within our temporal $(t > 0)$ universe, and it is also incorrectly interpreted as directional-independent, as can be seen from Eq. (5.3). It is, however, correctly treated special theory as a directional sensitive theory because a particle in motion follows its velocity vector. From this we see that the relativistic distance between two particles on the "same direction" can be shown as:

$$d_r = (v_1-v_2)\Delta t', \quad v_1 > v_2 \tag{5.9}$$

From this the faster particle has a section of time-gain relative to the slower one as given by:

$$\Delta t'' = (1-v_2/v_1)\Delta t', \quad v_1 > v_2 \tag{5.10}$$

For example, the distance between San Diego and New York City is:

$$d = c \, \Delta t' \tag{5.11}$$

and we have the following relationship:

$$\Delta t'' = (1 - v_2/v_1)d/c, \quad v_1 > v_2 \tag{5.12}$$

this represents a section of time-gain of a faster aircraft with respect to the slower aircraft between a distance d, but it is not a section of time-gain with respect to the universe since the time within an aircraft is the same as any subspace within our universe. Again, we see that we can change the relative arrival time between the two aircrafts, but we cannot change the time within our universe.

Although Einstein's special theory fails to legitimize within our temporal (t > 0) universe, yet his special theory had produced one of the most relevant energy equations as given by:

$$E = Mc^2 \tag{5.13}$$

where E is energy, M is mass, and c is the velocity of light. From this we see that our current science is limited by the speed of light. The fact is that Einstein's energy equation had given us the fundamental limit of energy; that is, mass and energy are equivalent. In this one of most famous and important equations that more than haft the world's population may have known it but many of them may not actually understand its physical significance. Anyway, this equation strictly speaking it is not consistent with a physically realizable axiom of our temporal (t > 0) universe since the special theory was developed from a non-physically realizable paradigm as I had shown. Yet the physical significance of Einstein's energy equation remains; energy and mass are equivalent.

5.5   A NEW MASS ENERGY EQUATION

It is trivial that Einstein's relativity equation can be written in a relativistic mass formula as given by:

$$M = M_0(1 - v^2/c^2)^{-1/2} \tag{5.14}$$

where M is the effective mass (or mass in motion), M_o is the rest mass, v is the velocity of the moving M, and c is the speed of light. In other words, the effective mass of a moving particle increases at the same amount with respect to the re-lativistic time window $\Delta t'$ (i.e., time dilation). Since Einstein's special theory was developed within an empty subspace, it has no induced gravitational field, as can be seen in Figure 5.11(a). Besides, it is not a physically realizable subspace paradigm, and it is trivial that empty space cannot support an induced gravitational field. In contrast, if it is situated within a temporal (t > 0) subspace shown in Figure 5.11(b), an induced gravitational field F is inherently attached with a temporal (t > 0) mass M(t) in motion.

(a)   (b)

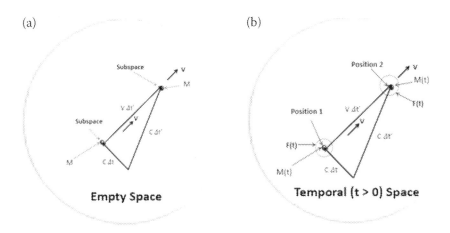

FIGURE 5.11 Shows the derivation of Einstein's special theory of relativity. (a) Shows within an empty subspace paradigm, (b) shows within a temporal (t > 0) space paradigm F(t) represents an induced gravitational field that is attached with a moving temporal (t > 0) mass M(t).

In the preceding we had rebutted the legitimacy of Einstein's special theory on the grounds of physical realizability, then on the same grounds his relativistic mass equation is in question. Although his energy equation provides a legitimate meaning (i.e., mass and energy are equivalent), it is the energy equation derived from a non-realizable relativistic equation. Since $E = Mc^2$ was developed on kinetic energy dynamics $E = (1/2) Mv^2$, it is logical to change the energy equation based on kinetic mass-energy dynamics as given by:

$$E = (1/2) \ Mc^2 \qquad (5.15)$$

which obeys the mass to energy dynamic, yet limited by speed of light. The mass-energy equation has an identical significance as Einstein's energy equation; mass and energy are equivalent, but it is more consistent with a physically realizable principle.

Since Einstein's special theory of relativity was derived on an empty space platform [i.e., Figure 5.11(a)], we had seen an induced gravitational field by mass M that had been ignored. As I had shown if the derivation is carried out within a temporal (t > 0) space shown in Figure 5.11(b), the total energy from a rest mass M should have been written as mass-energy plus an induced gravitational energy term as given by:

$$E'' \approx E + E_g \approx (1/2)Mc^2 + kM \qquad (5.16)$$

where $E = (1/2)Mc^2$ is the mass-energy conversion, $E_g = kM$ is the induced gravitational energy due to M, $k = (4/3)G_0 \ (r_0)^2$ a modifier gravitational constant, and E'' represents the total energy revervoir of a given mass M as exists within a temporal (t > 0) space.

From this we see that a temporal (t > 0) mass M(t) has existed within a temporal (t > 0) space prior to the big bang explosion. Since a temporal (t > 0) subspace has compacted with substances, we see that the induced gravitational field F(t) increases as M(t) increases. In view of total energy E" thermo-neuclei energy of E is much larger as compared with the induced gravitational energy E_g (i.e., $E \gg E_g$), it is evidently without the existence of the induced gravitational field by M(t), it will be very difficult to justify that the big bang explosion was started by mass M(t) itself [10]. From this we see that M (t) is a temporal (t > 0) substance that grows with time.

Although the big bang is capable of releasing a huge amount of energy from a thermo-nuclei explosion [i.e., $E = (1/2)Mc^2$] compared to the gravitational energy F_m (i.e., $E \gg E_m$), but without gravitational force to trigger the thermo-nuclei explosion, it is very difficult to convince us that big bang creation was ignited by itself, as most cosmologists had assumed. From this it must be a huge induced gravitational force that had ignited the big bang creation, but not by time as many consmologists had assumed.

Nevertheless, Eq. (5.16) is still a timeless (t = 0) equation, which cannot be directly implemented within a temporal (t > 0) space. In order for this equation to be implimented within a temporal (t > 0) subspace, we can transform this equation into a time-varying partial differential form as given by:

$$\frac{\partial E''(t)}{\partial t} \approx -\left[\left(\frac{1}{2}\right)c^2\frac{\partial M(t)}{\partial t} + K\frac{\partial M(t)}{\partial t}\right] \approx [\nabla\cdot[S(t) + S'(t)], \ t > 0 \quad (5.17)$$

where $k = (4/3)G_0 (r_0)^2$, $\nabla\cdot$ represents a divergent operator, S and S' are the respective thermo-nuclei and gravitational energy vectors which remain to be developed, and (t > 0) denotes that the equation is subjected to a temporal (t > 0) constraint.

This equation shows that right after the thermo-nuclei mass annihilation there were two major energies merging: thermo-nuclei and gravitational explosive energies. In other words, beside the thermo-nuclei energy as predicted by the mass-energy equation, there is a gravitational energy explosion by its own induced gravitational field. Thus, we see that part of the gravitational energy is an essential component for the creation of gravitation waves, as can be detected [11]. This indicates that an induced gravitational field had played a significant role for the big bang creation of our universe.

5.6   INDUCED GRAVITATIONAL FIELD

One of the esoteric features of Einstein's general theory is curving the time-space [3]; as John Wheeler had said: Space-time tells matter how to move; matter tells space-time how to curve. However, as I see it, it is time tells space how to curve but space does not tell time how to curve. In this I mean that it is time changes time-space, but not time-space that changes the time. The general theory was developed based on a Minkowski's space-time continuum where time is treated as an independent variable. However, from temporal (t > 0) universe standpoint, time and space coexist for which time is a dependent forward moving variable that moves at a constant pace. In other words, within our temporal universe, it is time that changes time-space, but not time-space curves the pace of time.

As we have accepted the origin of our universe was started from a big bang explosion within a temporal (t > 0) space [1,2], instead the creation within an empty space as normally assumed [10]. Then before big bang started a question may be asked: what triggered the explosion? As I had shown, it must be ignited by an intense convergent gravitational force created by a gigantic mass M(t) that eventually triggers the thermo-nuclei explosion. By this a question inquires under what physical condition will ignite the big band explosion. Since science is a law of approximation, I would assert that it must come from an extreme convergent gravitational force that had been induced by a temporal mass M(t), as shown in Figure 5.12.

In this we see that M(t) had been situated within a temporal (t > 0) space before the big bang started. Since temporal (t > 0) space is non-empty, it allows M(t) to create (i.e., store) its gravitational field (i.e., energy) induced by the existence of M(t). From this M(t) is capable of continually attracting substances added into its mass. In other words, M(t) behaves like a giant black hole (or it is a black hole) [11] that is constantly swallowing more and more substances. Over time, M(t) is getting heavier and heavier until its stored gravitational force reaches to a critical point that triggers the thermo-nuclei explosion of M(t). From this we see that big bang explosion must be ignited by huge gravitational pressure, instead of triggered by time as most cosmologists had assumed. For this big bang explosion must be due to extreme gravitational pressure of its own induced gravitational field that triggers the thermo-nuclei explosion of mass M(t), at least from our current physical standpoint.

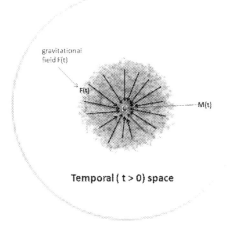

FIGURE 5.12 Shows a scenario well before the big bang explosion. Dark dot represents a point-singularity approximated gigantic temporal (t > 0) mass M(t), F(t) represents a huge gravitational field created by M(t), the arrows show a set of very intense convergent gravitational forces are applying on a mass M(t).

Nevertheless, it is the preexisting temporal (t > 0) space paradigm that provides all the possible temporal (t > 0) substances to continuously feed M(t) for the eventuality of self-explosion by its own weight. In other words, without the preexistent temporal (t > 0) space paradigm, we would not have a justifiable scientific reason that big bang theory had occurred. This is one of the many examples that show that a physically realizable solution comes from a physically realizable paradigm. Of this we see that big bang creation cannot get started within an empty space, as many cosmologists had assumed it was.

5.7 GRAVITATIONAL ENERGY

Since mass M(t) and its gravitational field are temporal (t > 0), the induced gravitational field coexists with M(t), as given by:

$$F(r; t) = G \ m \ M(t)/r^2 \tag{5.18}$$

where we see that induced gravitational force strength F(r; t) decreases inversely with the square law of distance r, G is a gravitational constant, and m represents an unit reference mass (i.e., points of interest) away from M(t), as illustrated in Figure 5.13.

With reference to the point of interest m, potential energy with respect to a unit m is assumed at a distance r away from a gigantic mass M(t) is given by:

$$E' = G_0 M(t)/r \tag{5.19}$$

where $G_0 = G \cdot m$ is a normalized gravitational constant. From this we see that the stored gravitational energy induced by M(t) is exponentially increasing as distance closes to mass M(t), as plotted in Figure 5.14.

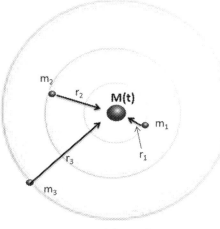

Temporal (t > 0) space

FIGURE 5.13   Shows units induced gravitational forces converge at mass M(t).

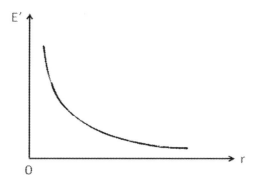

FIGURE 5.14 Shows the induced gravitational potential energy decreases rapidly as a function of distance r away from mass M(t).

Since the overall stored gravitational energy is around and deep into the mass M(t), we see that as M(t) reduces its mass, each reference m moves away with time when induced gravitational force loses its pull, as illustrated in Figure 5.15. The outward force acting on the point of interest m by the Newtonian second law can be written as:

$$f = ma \qquad (5.20)$$

where f is the acted force on m and a is the acceleration as given by:

$$a = G \ M(t)/r^2 \qquad (5.21)$$

This equation shows that acceleration on unit m is proportional to the inversed square law of distance r. In other words, the farther away from mass M(t) the

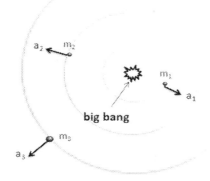

Temporal (t > 0) space

FIGURE 5.15 Illustrates a thermo-nuclei big bang hypothesis where the storage gravitational field releases its energy as the induced gravitational field rapidly shrinks with time as mass of M(t) annihilated due to a thermo-nucleoli explosion.

smaller the acceleration. While closer to M(t), m is anticipated to be accelerated even faster. For this, as stored gravitational energy shrinks rapidly, we anticipate a huge amount of outward gravitational force, due to the induced storage gravitational energy diverges as M(t) annihilates by thermo-nuclei explosion (i.e., big bang creation). From this we anticipate that a huge amount of stored gravitational energy together with a gigantic thermo-nuclei explosion will simultaneously release instantly right after the big bang started.

From this we see that without the thermo-nuclei mass annihilation there would be no such magnitude of gravitational waves that can be detected. Unlike the electromagnetic waves, gravitational waves are mostly longitudinal waves that dissipated quickly due to mass m in motion, as in contrast with a transversal electromagnetic wave. Thus, we see that it is extremely difficult to detect a gravitational wave that is due to mass or masses in motion within our temporal universe since longitudinal waves dissipate quickly within an absorptive time-space.

5.8   A NEW BIG BANG THEORY

Nevertheless, there are two dominant energy storages associated with any mass M(t) within a temporal (t > 0) space: gravitational storage energy E' and thermo-nuclei energy E, as given respectively by:

$$E' = G_0 M/r \tag{5.22}$$

$$E = (1/2)M \ c^2 \tag{5.23}$$

Since E' is the potential energy as referenced to point of interest m, the amount of total gravitational energy stored within the gravitational field induced by M(t) can be approximated by:

$$E'' \approx -G_0(4/3\pi)(r_0)^2 \tag{5.24}$$

This equation shows that stored gravitational energy E'' decreases as mass M annihilates; for this we see that a huge amount of gravitational energy releases right after M(t) has totally annihilated. In other words, an intense divergent gravitational shock wave releases almost simultaneously with thermo-nuclei explosion at the same moment, within the newly created expanding universe, as depicted in Figure 5.16. This shows that our universe is a temporal (t > 0) stochastic time-space that constantly changes naturally with time. And this is the nature of every temporal (t > 0) space where space changes naturally with time and space cannot change time.

Although these energy equations are timeless (t = 0), they can be reconfigured respectively into time-varying partial differential forms, as given by:

$$\frac{\partial E''(t)}{\partial t} \approx -k\frac{\partial M(t)}{\partial t} = [\nabla \cdot S'(t)], \quad t > 0 \tag{5.25}$$

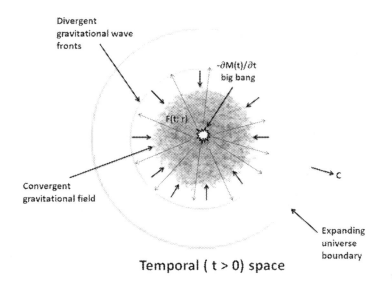

Divergent
gravitational wave
fronts

$-\partial M(t)/\partial t$
big bang

F(t; r)

Convergent
gravitational field

c

Expanding
universe
boundary

Temporal (t > 0) space

FIGURE 5.16   Shows a composited diagram of how our universe was created by a big bang explosion within a preexisting temporal (t > 0) space. The set of convergent arrows represents a shrinking gravitational field. A set of outward arrows represents a divergent gravitational wave due to rapid mass annihilation, and the boundary of our universe is expanding at the speed of light c due to a thermo-nuclei big bang creation.

$$\frac{\partial E(t)}{\partial t} \approx -\left(\frac{1}{2}\right)c^2\frac{\partial M(t)}{\partial t} = [\nabla \cdot S(t)], \quad t > 0 \tag{5.26}$$

where $k = (4/3)\pi\, G_0\,(r_0)^2$, $\nabla\cdot$ represents a divergent operator, S' and S are the respective gravitational and thermo-nuclei energy vectors, and (t > 0) denotes that equation is subjected to the temporal (t > 0) constraint. From this we see that right after the big bang explosion, two packages of divergent energies emerge, as illustrated in Figure 5.16. One is due to the thermo-nuclei explosion of M(t) and the other is due to gravitational waves, as can be seen in the diagram. Since a thermo-nuclei explosion is responsible for the big bang creation, the boundary of our universe expands at the speed of light. As for the gravitational wave, it is represented by a set of divergent arrows initiated by the big bang explosion. With this we see that a set of convergent arrows represents the collapsing gravitational field as mass of M(t) reduces.

Because every subspace within our universe is created by an amount of energy ΔE and a section of time Δt or (ΔE, Δt), we see that it is a necessary cost (i.e., a building block) to pay for every subspace creation, which includes the creation of our temporal (t > 0) universe itself. For instance, it took a huge amount of big bang energy ΔE to create our universe as given by:

$$\Delta E \approx (1/2)M\ c^2 \tag{5.27}$$

and a section of time

$$\Delta t \approx 14 \text{ million light years} \tag{5.28}$$

to create [1,2]. From this we see that every subspace changes with time. But our universe and its subspaces cannot change the speed of time, since time and subspace (i.e., substance) coexist. Thus, every subspace within our universe has the same time speed. Although as commonly believed that their relativistic time between subspaces may not be the same as based on Einstein's special theory of relativity, but I had already shown in the preceding. Einstein's special theory had failed to exist within our universe. Nevertheless, our universe as a whole runs at the same pace of time; if any subspace has a different time speed with respect to the time speed of our temporal (t > 0) universe, then the subspace cannot exist within our universe, which includes the timeless (t = 0) subspace. The fact is that those timeless and time-independent subspaces are virtual and non-physically realizable subspaces. For this, it is incorrect to assume those virtual and non-physically realizable spaces are "inaccessible subspaces" within our universe as some scientists do.

5.9   LEGITIMACY OF EINSTEIN'S GENERAL THEORY

As I had shown earlier that the empty subspace paradigm had hijacked the legitimacy of physical reality for decades, for which Einstein's general theory cannot be the exclusion since his general theory was also developed on the same empty timeless (t = 0) platform. Einstein's general theory was based on a past deterministic platform to predict a future consequence as given by [3]:

$$G_{\mu\nu} + \Lambda \ g_{\mu\nu} = (8\pi G/c^4) \ T_{uv} \tag{5.29}$$

where $G_{\mu\nu}$ is the Einstein tensor, $g_{\mu\nu}$ is the metric tensor, T_{uv} is the stress-energy tensor, Λ is the cosmological constant, G is the Newtonian constant of gravitation, and c is the speed of light in vacuum. But we see that his theory is a timeless (t = 0) point-singularity "deterministic" equation, instead of a non-deterministic principle. The fact is clear, since using deterministic consequences to predict a future event will be deterministic. But a future event is supposed to be non-deterministic or uncertain. In other words, the farther away from the present absolute certainty (i.e., t = 0) the more unpredictable a future consequence. From this we see that Einstein's general theory of relativity is a non-physically realizable theory within our universe since general theory predicts future events exactly. But future predictions are not supposed to be deterministic.

There are scores of examples that had used Einstein's general theory to predict the future outcome of solution; for example black holes [11], wormholes [12], time traveling [13], and others. Let me pick one of them to illustrate that prediction from general theory is virtual, fictitious, and hilarious as follows.

Let me first epitomize our temporal (t > 0) universe, as diagramed in Figure 5.17, in which we see that our universe started from a big bang creation and then changed

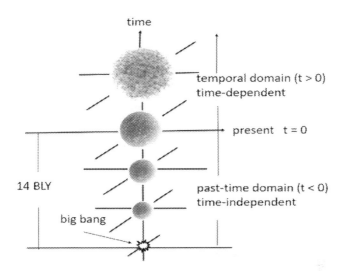

FIGURE 5.17 Shows our universe started from a big bang creation changes naturally with time. In this we see that the age of our universe is about 14 billion light years (BLY) old.

naturally with time. It shows that the age of our universe is about 14 billion light years (BLY) old, where the past-time domain (t < 0) represents a set of past-certainty universes (i.e., images or information) but without any physical substance and real time in it. The future-time domain (t > 0) shows a physical realizable but uncertain universe. From this we see that our universe changes naturally with time [i.e., temporal (t > 0)]. In other words, farther away from instant certainty the more uncertain the future prediction would be. This is also the reason that science is supposed to be approximated but not exact. Yet, time t = 0 represents an instant present moment of absolute physical reality of our universe, which has substances coexisting with time. But this present moment moves instantly to becoming a new present absolute certainty moment at a pace of time Δt (i.e., Δt → 0). And this is the summary of our temporal (t > 0) universe that I had described. From this we see that it is a mistake to treat our temporal (t > 0) universe as a four-dimensional space-time continuum as most scientists did. In other words, it is our universe that changes with time but not our universe that changes the time, as contrasted with Einstein's general theory of Eq. (5.29) describes (see Appendix E).

Yet, in view of the past-time domain (i.e., from big bang to present time t = 0) we see that the past universes were changing precisely with time. But past universes have no physical substances and have no real time. Yet, those past universes are a set of past information located at a specific time but without physical substance. From this mathematically speaking, we can treat them as a set of time-independent (virtual) universes, which is precisely where Einstein's general theory was developed from. And this is precisely why his general theory is deterministic, since certainty consequences predicted deterministic solution. But within our temporal (t > 0) universe every prediction is supposed to be approximated or non-deterministic.

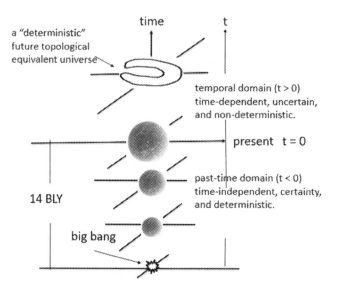

a "deterministic" future topological equivalent universe

time   t

temporal domain (t > 0) time-dependent, uncertain, and non-deterministic.

present   t = 0

past-time domain (t < 0) time-independent, certainty, and deterministic.

14 BLY

big bang

FIGURE 5.18 Shows a four-dimensional space-time continuum as most cosmologists and astrophysicists had assumed. Since within the future-time domain (i.e., t > 0) it is a universe that changes with time or a time-dependent space, from which we see that a four-dimensional space-time continuum is not a physically realizable paradigm.

Nevertheless, it must be the application of general theory, to my amazement, that had created a worldwide conspiracy. For instance, just pick one of many examples, as applied to wormhole time traveling. Let me start with the deterministic general theory of Eq. (5.29), from which it shows that it is possible to curve our universe to an equivalent topological shape, as illustrated in Figure 5.18. Then my question is what would be the price paid for curving our universe? Firstly, the general theory of relativity is a timeless (t = 0) theory; it has no price to pay in terms of time (i.e., Δt) to curve a space. Secondly, our universe has time, then how long will it take to curve our universe, as shown in the diagram? And this is precisely what the general theory had promised; it has no price to pay since Einstein's time is independent with space.

From this we see that an amazingly brilliant theory makes an impossible journey possible. But this only can happen if we are living within an empty universe, which does not need time to travel. On the other hand, if our universe is situated within a timeless (t = 0) space, then why do we bother to curve the universe since we can go anywhere instantly we wish. But unfortunately, this is not a timeless (t = 0) universe that we are living in.

Let us get back to the physically realizable temporal (t > 0) space of reality. My question is that how long (i.e., a section of time Δt) would it take to curve the universe and also how much energy of ΔE is needed? For simplicity, we can disregard the amount of information ΔI (i.e., or equivalently amount of entropy ΔS) to make it sufficient. From this we see that it would take a billion light years to curve our universe to the one shown in the diagram. In other words, every change within our temporal (t > 0) universe has a price to pay that is a section of time Δt and an amount of energy ΔE. Even though we assume that we have the price (i.e., Δt, ΔE)

to pay, but we are still unable to do it because our universe cannot curve our time-space. From this we see that Einstein's general theory is not a physically realizable theory since matter cannot curve time-space.

Nevertheless, it is possible to reconfigure Einstein's general theory to become a physically realizable principle, by imposing a temporal constraint on Eq. (5.29) as given by:

$$G_{\mu\nu} + \Lambda g_{\mu\nu} = (8\pi G/c^4)T_{uv}, \quad t > 0 \tag{5.30}$$

where $t > 0$ denotes that the equation is subjected by a temporal constraint for which any solution comes out from this equation will be temporal ($t > 0$) or physically realizable.

In summary, we have seen from Newtonian mechanics to Hamiltonian, to statistical, to wave mechanics, to relativistic and quantum mechanics are timeless ($t = 0$) mechanics. Although those timeless ($t = 0$) mechanics paved the way to our modern science, the basic empty space paradigm has not changed. For this it has produced a number of unthinkable virtual timeless ($t = 0$) solutions that are causing a worldwide scientific conspiracy. Regardless of it is inadvertent or not, it is our responsibility to change it back to physically realizable science. Otherwise, we will be continually trapped within the wonderland of timeless ($t = 0$) science, which does not need to pay any price (i.e., Δt, ΔE). But unfortunately, within our temporal ($t > 0$) universe, everything needs a price to pay a section of time Δt and an amount of energy ΔE and it is not free. For which we see that, a four-dimensional space-time continuum as depicted in Figure 5.19, is not a physically realizable paradigm. Since time-independent space is a zero-summed energy

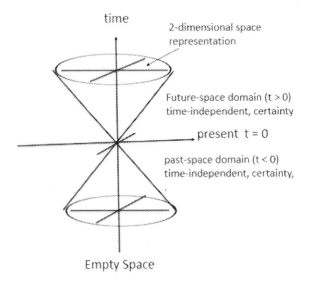

FIGURE 5.19 Shows a four-dimensional space-time continuum that most theoretical physicists had used. It is zero-summed energy subspace that most of the cosmologists and astrophysicists had used.

subspace that allows anti-particle to exist [15], but anti-matter cannot exist within our temporal (t > 0) universe.

5.10 CURSE OF SIMULATION

Theotherical physicists used a very convincing simulation, yet simulation is as virtual as mathematics. Without support by a physical realizable paradigm, their simulated result would very likely be unrealizable as many analytical solutions are. In other words, it is not how fancy a simulation is, it is the physically realizable paradigm that gives us physically realizable solutions. Since there are scores of fictitious simulations, let me pick up one of the most obvious simulated examples that I have seen many times and as well had published in well-known textbooks [14], which is using a simple Pythagorean theorem to simulate time-dilation of special relativity.

For example, the simulation starts with two sets of reflecting mirrors: one is stationary and the other is moving at a constant velocity toward the right-hand side. If we equip these two platforms, each has a clock with it and provide a light beam that is to be used for reflecting back and forth between two mirrors. We further assume each reflection is synchronized with the clocks for one click, as depicted in Figure 5.20(a). Since it is possible for us to split a light beam into two by means of a

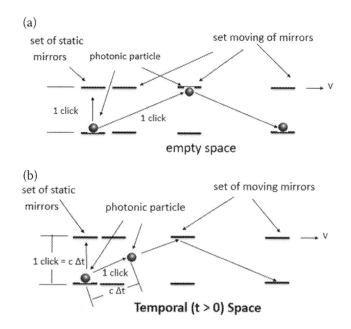

FIGURE 5.20 Shows that two identical perfect clocks, one is situated within a stationary platform and the other is carried by a moving platform. For simplicity, I have used photonic particles to identify the location, v is the velocity of the moving platform, c is the speed of light, and Δt is a section of time. (a) Shows a simulation that is performed within an empty space. (b) Shows the same simulation but it is performed within a temporal (t > 0) subspace.

beam splitter at the lower mirror, from this we see that one light beam within the stationary platform is allowed one click per reflection, and the tilted beam as reaching the moving platform is synchronized with the reflecting beam within the stationary mirrors, that is one reflected light per click as the clock moves, which can be seen in the figure. From this we see that as the clock travels with the moving platform, it seems the clock clicks slower than the stationary clock, and it is because the tilted light beam travels a longer distance per click. This simulated configuration probably is one of the most convincing pieces of evidence for millions of scientists to agree that Einstein's prediction is legitimized. That time indeed moves slower within a moving subspace as relative to a static subspace. And this is one of many other simulated results that have convinced scores of scientists.

Nevertheless, if we submerge the same scenario within a temporal $(t > 0)$ subspace, as shown in Figure 5.20(b), where the real time within our subspace cannot be ignored. Quickly we found out that the clock that situated within the moving subspace is clicking at the same pace as the other clock that is situated within a stationary platform. From this we see that Einstein's special theory is in question.

Furthermore, let me quote one of the very convincing experiments as reported by Hafele and Keating [16]. In which an atomic clock flew twice around the world, first eastward, then westward, and compared with another atomic clock against other that remained behind at the United States Naval Laboratory. When they reunited, these clocks were found to disagree with one another; they had concluded that their experiment was consistent with Einstein's special relativity. Nevertheless, every physical aspect within our universe changes with time, and the atomic clocks that Hafele–Keating had used cannot be the exception. Since the reunited clocks gave different readings, it is trivial that, physically these reunited clocks cannot coexist, since neither a past-clock nor a future-clock can be seen at present physical reality. For example, we cannot coexist yesterday of us or even tomorrow of us. Then should we trust the conclusive evident of Hafele-Keating experiment or the physically realizable interpretation of science? For which if one performs Hafele-Keating once or more times, I am certained their result would not be as conclusive as relativity theory anticipated.

Aside from the physical fact that we cannot change time, there are scores of fantasy hypothetical sciences, since Einstein's theory and Schrödinger's principle introduced a century ago. And this is precisely who we owe to the physical realizable science; otherwise we will continually be trapped within the fantasy timeless $(t = 0)$ wonderland of virtual science. For this I have seen that physically realizable science has been overshadowed by mathematical simulation, since science needs mathematical simulation. But again, mathematical simulation is not science, from which I have seen scores of virtual simulated solutions. That had already created a score of worldwide fantasy sciences, ranging from non-physically realizable qubit computing to absurd relativistic wormhole time traveling.

5.11 REMARKS

Einstein's special and general theories were the most intriguing esoteric principles in the last century of modern physics, and they were and still are the cornerstones of modern space-time science. Yet, I have recently found that his theories were developed on an empty timeless (t = 0) subspace platform for which his special theory and general theory are timeless (t = 0) or time independent. Since within our universe every subspace is temporal (t > 0), it is a time-dependent variable time-space, in contrast with most scientists who assumed time is an independent variable. In view of temporal (t > 0) universe I have shown that time can only move forward at a constant speed; yet time cannot be delayed or moved ahead of its own pace by subspace, since time and space coexist, which is the nature of time-space.

Since the special theory of relativity has been treated as a relativistic-directional independent as the equation provided, I had shown that Einstein's special theory is in fact a relativistic-directional dependent principle, even under the assumption the special theory of relativity is physically realizable. I had also shown it is time-space that changes with time, but not time-space curves (i.e., changes) time. As in contrast with his general theory has stated, it is the matter that curves the space-time since the theory was derived from a time-independent Minkowski space.

Nevertheless, one of the most important legacies of Einstein's special theory must be the derivation of his famous energy equation. Again, the root of this equation was also developed from the same empty subspace platform, as many of us had unintentionally been using it. From this I have found a viable induced gravitational energy term that should have had added with his energy equation. This newly found gravitational energy could be an important consequence in view of the big bang creation, by which normally has assumed the big bang explosion was ignited by time within an empty space. From this I have shown that besides a huge thermo-nuclei big bang explosion, an intense gravitation wave was simultaneously released within the explosion. Since an indued gravitational field had never been a part that triggers the big bang creation, the reason must be big bang creation had already been assumed within an empty space, by most cosmologists well before the big bang postulation.

Time is a dependent-forward variable that I have shown is the nature of our temporal (t > 0) space. From this I have shown that it is the moment of t = 0 that divides the virtual and physical realities. In other words, past certainty is a time-independent virtual reality subspace, while future uncertainty is a time dependent physical reality subspace. From this we see that practically all the laws, principles, theories, and paradoxes were developed from the past (i.e., t < 0) consequences to predict the future outcomes but with deterministic solutions. This includes Einstein's relativity theories and Schrödinger's quantum mechanics are time-independent deterministic theories and principles. But our universe is temporal (t > 0); strictly speaking, all the deterministic solutions cannot exist within our temporal (t > 0) universe, since our universe is a time-dependent subspace. Yet, those deterministic laws, principles, and theories are propelling us to a new temporal (t > 0) realm of science, in which we see that it is science that changes with time but not science that changes the time.

Nevertheless, I have shown that Einstein's special and general theories are non-physically realizable within our temporal (t > 0) universe; they are as virtual as mathematics is since they were developed from a timeless (t = 0) subspace platform that does not exist within our universe. Since a timeless (t = 0) subspace paradigm has undermined the physical reality of science, it is time for us to switch to temporal (t > 0) science instead of continually trapping ourselves within the virtual timeless (t = 0) wonderland of science. Finally, a remark is that physical reality is the deciding fact of science and should not let rigorous mathematics decide the fate of reality, since science is supposed to be approximated.

REFERENCES

1. F. T. S. Yu, "Time: The Enigma of Space", *Asian Journal of Physics*, 26 (3): 143–158, 2017.
2. F. T. S. Yu, "From Relativity to Discovery of Temporal (t > 0) Universe", *Origin of Temporal (t > 0) Universe: Correcting with Relativity, Entropy, Communication and Quantum Mechanics*, Chapter 1, CRC Press, New York, 1–26, 2019. New York.
3. A. Einstein, *Relativity, the Special and General Theory*, Crown Publishers, New York, 1961.
4. R. Zimmerman, *The Universe in a Mirror: The Saga of the Hubble Space Telescope*, Princeton Press, Princeton, NJ, 2016.
5. F. T. S. Yu, "Schrödinger's Cat and His Timeless (t = 0) Quantum World", *Origin of Temporal (t > 0) Universe: Correcting with Relativity, Entropy, Communication and Quantum Mechanics*, Chapter 5, CRC Press, New York, 81–97, 2019. New York.
6. K. Fredenhagen, "On the Existence of Antiparticles", *Communications in Mathematical Physics*, 79: 141–151, 1981.
7. O., Belkind, "Newton's Conceptual Argument for Absolute Space", *International Studies in the Philosophy of Science*, 21(3): 271–293, 2007.
8. T. D. Lee and C. N. Yang, "Question of Parity Conservation in Weak Interactions". *Physical Review*, 104 (1): 254–258, 1956.
9. B. P. Abbott, et al. (LIGO Scientific Collaboration & Virgo Collaboration), "GW170817: Observation of Gravitational Waves from a Binary Neutron Star Inspiral", *Physical Review Letters*, 119 (16), 2017.
10. M. Bartrusiok and V. A. Rubakov, *Introduction to the Theory of the Early Universe: Hot Big Bang Theory*, World Scientific Publishing, Princeton, NJ, 2011.
11. M. Bartrusiok, *Black Hole*, Yale University Press, New Haven, CT, 2015.
12. M. S. Morris and K. S. Thorne, "Wormholes in Spacetime and Their Use for Interstellar Travel: A Tool for Teaching General Relativity", *American Journal of Physics*, 56 (5), 395–412, 1988.
13. M. Visser, "From Wormhole to Time Machine: Comments on Hawking's Chronology Protection Conjecture", *Physical Review D*, 47 (2): 554–565, 1993.
14. 5. G. O. Abell, D. Morrison, and S. C. Wolff, *Exploration of the Universe*, 5th ed., Saunders College Publishing, New York, pp. 223, 1987.
15. P. A. M. Dirac, "The Quantum Theory of the Electron". *Proceedings of the Royal Society A*, 117 (778): 610–624, 1928.
16. J. C. Hafele and R. E. Keating, "Around-the-World Atomic Clocks: Observed Relativistic Time Gains", *Science*, 177 (4044): 168–170, July 14, 1972.

6 Nature of Entropy: An Energy Degradation Principle

In this chapter I shall begin with the nature of the temporal (t > 0) universe, since every subspace or substance has to be temporal (t > 0); otherwise it cannot exist within our universe. In other words, within our universe every law, principle, and theory has to be temporal (t > 0), which includes Boltzmann's entropy. Since entropy is always an intriguing subject, I shall reason that entropy increases naturally with time. This is primarily due to the expansion of our universe at the speed of light, of which the energy that created our universe is conserved. Since an entropy increase can be determined by energy degradation within our universe, as I shall show, this is due to the expanding universe that causes energy degradation where energy is conserved within our universe. For this our universe will never be actually totally disappeared, except at the point of infinity (i.e., t → ∞). Although the nature of entropy had indicated to us that time is dependent space, we opted to accept time as an independent variable for centuries. There was no other way to hypothesize a new hypothesis without using past certainty (i.e., information), yet deterministic principles produce exact solutions. The irony is that all the laws and theories are exact, for which scores of their solutions were as virtual as mathematics is, although we need mathematics. In short, it seems to me that we have lost our logical independent thinking and have opted to accept the approval of others. But as we look back all non-physically realizable principles and theories, this must be the right moment for us to look into the physically realizable paradigm. Otherwise, we will forever be trapped within the timeless (t = 0) wonderland of fantasy science, which does not have to pay a price (i.e., Δt, ΔE).

6.1 LAW OF NATURE

Let me begin with the nature of our temporal (t > 0) universe, the law of physical realizability, as depicted by Figure 6.1.

From this we see that our universe is not a time-independent space as most scientists had assumed. But in fact our universe is a time inter-dependent space [1,2], where absolute certainty exists if and only if at a present moment (i.e., t = 0) as epitomized in Figure 6.2.

From this we see that that present certainty divides the past time-independent but virtual certainties, and future to be physically realizable time-dependent but uncertain universe. Since the present moment (t = 0) of our universe is an absolutely certainty, this moment of physical reality will instantly move ahead to become the

DOI: 10.1201/9781003271505-6

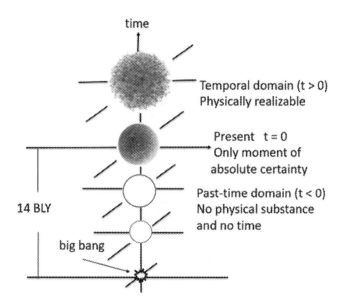

FIGURE 6.1 Shows that our universe started from a big bang creation about 14 billion light years (BLY) ago. Past-time domain (t < 0) represents certain virtual past images at precise locations, but have no physical substance and time. Future-time domain (t > 0) represents a physically realizable uncertainty universe, yet t = 0 represents the present moment of absolute physical certainty.

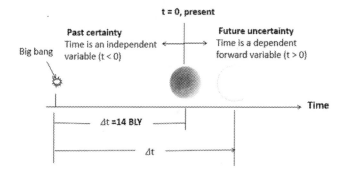

FIGURE 6.2 Shows present moment (t = 0) is the only absolute moment of truth. All the others are virtual without physical substance and no time. From this we see that the past is time independent of virtual certainty and the future will be a physically realizable time-dependent universe but uncertain.

next instant moment of reality and so on. For this, all the past and future are virtual realities and have no physical substance and real time is associated with them. Even though a future prediction is virtual, momentarily it will be physically realizable with degrees of uncertainty. As for the past, the certainty consequences of past realities but without physical substance. Yet the future is always a time-dependent physically realizable consequence.

Nevertheless, every subspace, no matter how small it is, within our universe was created by a section of time Δt and an amount of energy ΔE, and the relationship is given by:

$$\Delta t \; \Delta E = (\tfrac{1}{2}) m(t) c^2 \qquad (6.1)$$

This relationship represents a necessarily price or building block of our universe, but not sufficient yet to be actually created. From this we see that every substance or subspace within our universe is an energy conservation isolated subspace, where m is mass and c is the velocity of light. From which we see that, every isolated subspace within our universe is an energy conserved subspace.

Notice that this equation is somewhat different from Heisenberg's uncertainty principle [3] as from a physical consistence standpoint. Heisenberg's principle is an observational uncertainty principle, which is independent with time. As in contrast with the implication of Eq. (6.1), it is a time-dependent principle, since $m(t)$ changes naturally with time. In other words, we see that the pace of time within our universe cannot be changed. And this is the major issue why the uncertainty principle has to change with time, because it is profoundly connected with the creation of our universe. From this we see that energy created our universe is constantly degrading with time, since the big bang creation.

Similarly, within the quantum regime we have the uncertainty relationship as given by:

$$\Delta t \; \Delta E = h \qquad (6.2)$$

where energy is conserved, h is Planck's constant, Δt and ΔE are the variational discrepancies that change with time, $E = h\nu$ is a "quanta" of light or a quantum-leap package of energy, and ν is the quantum leap frequency. Nevertheless, a quantum uncertainty relationship can also be shown by an equivalently form as given by:

$$\Delta t \; \Delta \nu = 1 \qquad (6.3)$$

where $\Delta \nu$ is the corresponding bandwidth. Nonetheless, the nature of temporal $(t > 0)$ is that present moment $(t = 0)$, which is the only absolute physical certainty. But this moment of truth will instantly move ahead to become the next absolute certainty and so on with the pace of time. From this we see that the past and future are as virtual as mathematics is. Nevertheless, the past was the consequential certainty memories and the future will be the uncertain but physically realizable consequence.

6.2   NATURE OF UNCERTAINTY

Since our universe is an expanding temporal $(t > 0)$ stochastic bounded space, every subspace within the universe changes with time. From this we see that science is approximated. In other words, science cannot be absolutely or deterministically hypothesized. And this is precisely the reason why deterministic law, principle, and theory are in question.

Nevertheless, energy has different forms, as from mass annihilation principle, M is equivalent to a total energy since energy is conserved by $E = (\frac{1}{2}) Mc^2$. But from a momentum conservation standpoint, mass m in motion is equivalent to a total energy $E = (\frac{1}{2}) mv^2$, since momentum conservation (i. e., p = mv) is equivalent to kinetic energy. For this I had assumed that mass is self-annihilated.

However, as within the quantum theory regime, each quantum state energy is equivalent to a quantum leap energy of $E = hv$, where v is the quantum leap frequency.

Since our universe expands with time, entropy increases naturally with time [1,2], by virtue of the uncertainty principle, and their energy conservation can be written in the following forms:

From mass annihilation standpoint we have:

$$\Delta t \; \Delta E \geq (\tfrac{1}{2}) Mc^2 \qquad (6.4)$$

From momentum conservation consideration we have:

$$\Delta t \; \Delta E \geq (\tfrac{1}{2}) mv^2 \qquad (6.5)$$

For quantum leap energy conservation, we have:

$$\Delta t \; \Delta E \geq h \qquad (6.6)$$

From this we see that all the uncertainty principles change naturally with time since Δt and ΔE coexists with time, and that includes Boltzmann's entropy theory as given by [4];

$$S(t) = -k \; \ln \; p(t), \quad t > 0 \qquad (6.7)$$

where t > 0 denotes equation is subjected to temporal (t > 0) constraint, k is the Boltzmann constant, ln is the natural log, and p(t) is a temporal (t > 0) probability function decreases naturally with time. Since Boltzmann's entropy is measure of uncertainty that changes with time, it represents a necessary cost in terms of energy degradation. In other words, the more energy degraded within a subspace the higher the amount of entropy increases within the subspace. In other words, the more ambiguous or uncertain the subspace is supposed to be. In this we see that the higher entropy subspace, the higher the information content, since information is measured within a subspace can in principle be written as [5,6]:

$$I(t) = -\log_2 \; p(t), \quad t > 0 \qquad (6.8)$$

in bits. From this we see that entropy and information in principle can be traded [6,7]. But remember an amount of entropy is equivalent to an amount of information (e.g., in bits), but they are not equal.

6.3   AN ENERGY CONSERVATION UNIVERSAL

Nevertheless, Boltzmann's entropy is a temporal $(t > 0)$ theory as given by:

$$S(t) = -k \ln p(t), \quad t > 0 \tag{6.9}$$

This means that the amount of entropy within an isolated subspace increases naturally with the pace of time (i.e., the law of nature), since $p(t)$ is a monotonously decreasing function of time. Similarly, an equivalent amount of information also increases with time, since every bit of information in reality created by a section of time Δt and an amount of energy ΔE, but ΔE degraded naturally with the pace of time.

Since entropy within an isolated space (or substance) increases with the degradation of energy within the subspace, this is dependent upon an amount of energy degradation, as depicted In Figure 6.3. From this an energy-degrading temporal $(t > 0)$ probability of $p(t)$ is written by:

$$p(\Delta t) = E(t - \Delta t)/E_0 = 1 - \alpha(t)/E_0, \, t \geq 0 \tag{6.10}$$

where E_0 represents a total energy within an isolated subspace, $E(t - \Delta t)$ is a monotonous energy-degrading function of t (or Δt) as depicted in the figure, and $\alpha(t)$ is the rate of energy degradation with time as given by:

$$\alpha(t) = -dE(t)/dt, \quad t > O \tag{6.11}$$

One of the important indications of Boltzmann's entropy is that time cannot be treated as an independent variable with space; otherwise entropy within an isolated subspace cannot increase with the pace of time. And this was precisely a crucial indication that our space is a time-dependent space. Yet, we had missed this

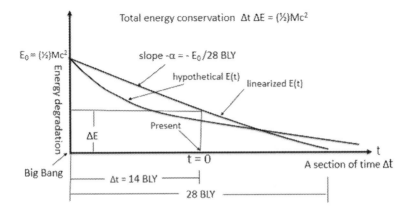

FIGURE 6.3 Depicts a universal energy degradation diagram, $E(t)$ is a monotonously decreasing function of time started from big bang explosion to the current present state of absolute certainty and beyond to future uncertainty consequence.

important implication for over a century and had persistently opted to treat time as an independent variable with space.

As we refer to the energy degradation curve depicted in Figure 6.3, we see that $\alpha(t)$ increases with time as our universe expands at the speed of light. Since $E(t)$ degrades monotonically with t (or Δt), it tells us that the amount of energy created our universe will eventually degraded as time approaches to infinity (i.e., $t \rightarrow \infty$). Since $E(t)$ is a nonlinearly monotonous decreasing with time, we see that the energy that created our universe will be eventually dissipated within a greater temporal $(t > 0)$ space that our universe was embedded in. But our temporal $(t > 0)$ universe is a subspace of a greater temporal $(t > 0)$ space, and energy [i.e., $E = (\frac{1}{2})Mc^2$] that crested our universe is conserved within our expanding universe. All the degraded energy will eventually be dissipated back within the greater temporal $(t > 0)$ space. Of that we see that our universe is a temporal $(t > 0)$ subspace within a greater temporal $(t > 0)$ space. This is similar to an expanding water wavelet on the surface of a water pond. Eventually it will disappear within the surface of the pond.

6.4   AN ENERGY DEGRADATION UNIVERSE

Nevertheless, the rate of degradation [i.e., $\alpha(t)$] in principle can be determined with reference to the expansion of our universe, as shown in the Figure 6.4.

In this we see that our universe it is an energy degrading temporal (t > o) universe. And this is the reason why entropy increases naturally with time, since our universe is a bounded subspace that expands at the speed of light. In short, our universe is a dynamic stochastic subspace that changes in volume with the pace of time. In view of Figure 6.4 it tells us that our universe has transformed from a big bang explosion at $t = -14$ billion years (BLY) ago. That was from a relatively small volume of v_0, to the current size at the speed of time. For this the rate energy degradation within our universe can be approximated by:

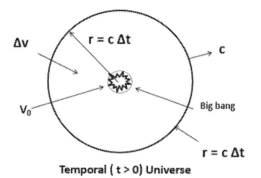

Temporal (t > 0) Universe

FIGURE 6.4   Shows the law of energy degradation within our universe. In this it shows that energy degraded naturally with time, as our universe expands at the speed of light. Where v_0 is the initial volume of mass M, Δv is the net volume gain with a section of time Δt, c is the velocity of light, and r is the radial distance.

$$\alpha(\Delta t) \approx v(\Delta t)/v_0 \qquad (6.12)$$

where $v(\Delta t)$ is the net volume gain and v_0 is the volume or the size of mass M before the explosion. Since the net volume gain is equal to the expanded volume minus the size of mass M, for this the net gain in volume Δv can be approximated by:

$$\Delta v = v(\Delta t) - v_0 \approx (4/3)\pi(c\ \Delta t)^3 \qquad (6.13)$$

since $v_0 \ll v(\Delta t)$]. In view of the hypothesized nonlinear energy degradation curve, as shown in Figure 6.3, we see that energy degradation within our universe will run to an infinitely long time (i.e., $t \rightarrow \infty$), as the postulated curve shows in the figure. Since it is highly non-linear energy degradation curve, this provides us with a complicated issue to tackle the problem as I shall address. Nevertheless, we see that right after the big bang explosion, energy degradation within our universe at that moment was anticipated to be very fast; it could had been within a fraction of a second as I hypothesized.

But, as soon our universe had grown bigger and larger at the speed of light, the net gain in volume slowdown is shown in Figure 6.5. It shows that as our universe had expanded to the current size, the net gain in volume for a given section of time Δt is compared for a smaller-based volume v_0 is relatively small. The rate of energy degradation is slowing down somewhat and eventually will taper off. But it will never be diminishing to zero. In other words, the total energy that created our universe will eventually degrade or dissipate into the greater temporal ($t > 0$) space that our universe had embedded in.

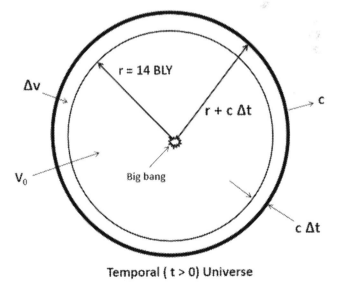

Temporal (t > 0) Universe

FIGURE 6.5 Dramatizes that net gain in volume as compared with the base volume v_0 becomes larger. From this we see that relative gain in volume (i.e., $\Delta v/v_0$) is slowing now as time moves on.

Yet, for the sake of simplicity illustration, I had linearized the energy degradation curve somewhat for convenient demonstration, as can be seen in preceding Figure 6.3, where I had assumed the total energy E_0 within our universe had been totally degraded at $\Delta t = 28$ BLY after the big bang explosion as given by:

$$E(\Delta t = 28 \ BLY) = E(t = 14 \ BLY) = 0 \qquad (6.14)$$

Yet, as we referred back to Figure 6.5, we can visualize that the net universal volume gain is equal to the volume of our universe after a section of time Δt minus the volume of our present universe. Since the volume of our present universe is given by:

$$v_0 = (4/3) \ \pi \ [(28 \ BLY)c]^3 \qquad (6.15)$$

And the volume after a section of time Δt is:

$$v = (4/3) \ \pi \ [(28 \ BLY)c + c \ \Delta t]^3 \qquad (6.16)$$

For this we have a net volume gain of:

$$\Delta v = (v - v_0) = (4/3) \ \pi \ (c \ \Delta t)^3 [(28 \ BLY + \Delta t)^3 - (28 \ BLY)^3] \qquad (6.17)$$

Thus, rate of energy degradation can be approximately by:

$$\alpha(\Delta t) \approx \Delta v/v_0 = 3\Delta t/28 \ BLY + 3(\Delta t)^2/(28 \ BLY)^2 + (\Delta t)^3/(28 \ BLY)^3 \qquad (6.18)$$

Which can be further reduced to:

$$\alpha(\Delta t) \approx 3\Delta t/28 \ BLY \qquad (6.19)$$

Then the corresponding energy degrading probability as started from big bang (i.e., $t = -14$ BLY) to another 14 BLY beyond the current moment $t = 0$, is a total of 28 BLY (i.e., $\Delta t = 28$ BLY), which can be shown as:

$$p(\Delta t)|_{\Delta t \leq 28 \ BLY} = [1-\alpha(\Delta t)] = 1-[\Delta v(t)/28 \ BLY]^3 \qquad (6.20)$$

From this we see that our universe had started from a very low entropy state as given by:

$$S(t = -14 \ BLY) = -k \ \ln \ p(t = -14 \ BLY) = 0 \qquad (6.21)$$

which was 14 BLY ago. Since then, entropy had been increasing with time. For this we see that there is an equivalent amount of energy that has been persistently degraded due to the increase in entropy.

But as we look back to the linearized energy degradation curve in Figure 6.3, we see that the total amount of entropy increased will be infinitely large, when all the energy within our universe has been totally degraded. Of this energy degrading probability for $\Delta t = 28$ BLY, this is at $t = 14$ BLY after the present moment $t = 0$ is given by:

$$p(\Delta t)|_{\Delta t=28 \text{ BLY}} = [1-\alpha(t)] = 0 \qquad (6.22)$$

This shows that an infinite amount of entropy had increased. This means that that life of our universe is terminated at $t = 14$ BLY (i.e., a total life time of 28 BLY), based on the linearized energy degradation hypothesized as shown. Since at the end of our universe, its entropy has increased to infinitely large as given by:

$$S(\Delta t = 28 \text{ BLY}) = -k \ \ln \ p(\Delta t = 28 \text{ BLY}) = \infty \qquad (6.23)$$

This means that total energy within our universe has been completely degraded or dissipated within a large temporal $t > 0$ space that our universe had embedded in.

Yet, linearized assumption is for simple demonstration, it is by no mean be precise. In view of the hypothetical degradation curve, we see that the undegraded energy within our universe will eventually degrade at the point of infinity, although the rate of energy degradation will be continuously slowing down at a much lower rate. This is very similar to a water wavelet that propagates on a still water pond, from which we see that our universe will eventually dissipate within a greater temporal space as time approaches to infinity (i.e., $\Delta t \rightarrow \infty$). In other words, all the energy that had created our universe will eventually degrade or give it back to the greater temporal space that our universe had created from. Since our universe is an energy conserved subspace obeys the second law of thermodynamics, it is a good reason for us to connect the degraded energy with the dark matter that remains to be seen?

6.5   ESSENCE OF BOLTZMANN'S ENTROPY

Firstly, did time exist because of the existence of us? In other words, are you a part of the universe or is universe a part of you? As a scientist he/her may tell us you are a part of the universe. However, from a philosopher's standpoint he/she may tell you it can be both ways. Nevertheless, if the question is changed to, are you a part of the human species or is the human species a part of you? Then the scientist and philosopher may both give you the same answer: we are part of the human species. Since we are dealing with science, everything should be based on an objective physical reality standpoint. As what I have shown in the preceding chapters that time and space coexist or are interdependent within our universe since our universe had to be created within a greater temporal ($t > 0$) space, otherwise our universe had no physical justification that can be created, as from a temporal ($t > 0$) exclusive principle standpoint. In this we see that everything changes with time, but we can neither change nor stop time. In other words, time is the only variable that moves at a constant pace coexisting within every subspace. For this it is space that changes naturally with time.

From this we see that Boltzmann's entropy is a relative measure based on energy degradation probability within a subspace, although it gives in energy units. For example, we have a small mass annihilation, and then we see that entropy will increase more rapidly as compared for a larger mass annihilation, since it has less available undegraded energy within the subspace to be degraded. And this is precisely the reason that a smaller particle is more difficult to predict or detect. For instance, it will take billions and billions of light years for our universe to exhaust its undegraded energy to reach its highest entropy statute. In this we see that Boltzmann's entropy is a measure of relative units of complexity but not actually represent the physical reality of the subspace. For example, Boltzmann's entropy is a one-side bounded equation from 0 to infinite, as given by:

$$S = -k \ln p \qquad (6.24)$$

But it does not give us the physical cost in terms of a section of time Δt and an amount of energy needed (i.e., Δt, ΔE). From this we see that, it is the total energy within an isolated subspace that determines a section of time Δt with an amount of energy ΔE is required. This is an equivalent amount of undegraded energy that can be degraded within the subspace. In this we see that the remaining amount of undegraded energy $\Delta E'$ is conserved, where $\Delta t' \, \Delta E' = E_0'$ is the total undegraded energy within the subspace, as can be seen in Figure 6.6. In other words, energy within an isolated subspace cannot be destroyed but can be changed or transformed from undegraded to degraded, since total energy is conserved within an isolated subspace.

6.6 ENTROPY BEYOND PRESENT CERTAINTY

Since Boltzmann's entropy is a temporal $(t > 0)$ principle, this is a measure of uncertainty of an isolated subspace, which is equivalent to the complexity of an

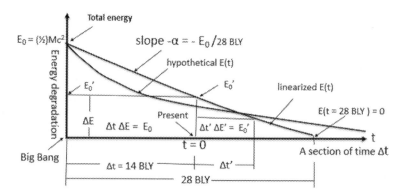

FIGURE 6.6 Shows energy degradation of our universe beyond the present moment $t = 0$. E_0' represents the amount of undegraded energy left which is available for degradation within our universe, under the linearize assumption. Where $E_0' = \Delta t' \, \Delta E'$ is conserved.

isolated subspace. But once the uncertain subspace is physically realized (i.e., exists physical reality), the uncertainty measure of the subspace becomes certain. This means that entropy of the subspace is the lowest (i.e., $S = 0$) although the subspace under entropy complexity measure is the highest. And this is precisely why entropy increases naturally with time, since our universe is an isolated temporal ($t > 0$) stochastic universe, where its boundary expands at the speed of light. Our universe was started from a very small singular approximated mass of gigantic weight that had exploded to create our universe. Where the size of the universe is continuous enlarging, this makes the universe more complex (i.e., in spatial term) at every expansion. This shows that entropy increases are in terms of complexity measure of subspace. From this we see that the present moment (i.e., $t = 0$) is the only moment of absolute certainty as from the temporally ($t > 0$) standpoint. This is different from spatial complexity where our universe entropy is the highest at the current moment, as compared to the past (i.e., $t < 0$). For this, certainty at the present moment is by no means back to the singularity assumption.

Nevertheless, the amount of degraded energy within our universe is derived from a section of $\Delta t = 14$ BLY is given by: $\Delta E' = E_0 - E_0'$, as depicted in Figure 6.6, which I repeat here for convenience. We see that an amount of degraded energy $\Delta E'$ is equivalent to an amount of entropy increase started from the big bang explosion to the present moment (i.e., $t = 0$). But $\Delta E = E_0 - \Delta E'$ is the remaining undegraded energy within our universe that can be degraded. Again, we note that present moment $t = 0$ is one and only one absolute certainty state. In other words, it is impossible to have two or more present absolute subspaces (e.g., current universe) simultaneously exist within our universe. From this we see that, Schrödinger's fundamental principle of super position fails to exist, as from the entropy theory standpoint.

For this remaining amount of energy $\Delta E' = E_0'$ will continuously degrade naturally with time, as our universe is expanding at the speed of light. We see that energy is conserved within our universe as given by:

$$\Delta E \times (\Delta t = 14 \ \text{BLY}) = E_0 = (\tfrac{1}{2})Mc^2 \tag{6.25}$$

But entropy increased between the big bang explosion to the present moment $t = 0$) is the uncertainty projection, yet entropy will continuously increase beyond the present moment to another 14 BLY based on linear projection, which I had assumed our universe to be terminated as shown in the diagram. But this is by no mean a physical reality, as I had noted in the preceding.

As we have accepted the current moment ($t = 0$) is the only moment of absolute certainty, we have seen our universe has been continually transforming with time from a physical reality to the other, and so far it has no end in sight. In other words, it has never been an absolute predictable universe. And this is precisely entropy within our universe at the present moment ($t = 0$) is lowest [i.e., $S(t = 0) = 0$]. But for any further hypothetical prediction beyond the present moment, entropy will increase. Energy conservation within our universe is based on the remaining un-degraded energy within the current universe E_0', as depicted in the diagram. This illustrates that entropy will continue to increase naturally from the present moment with time. This shows that an expanded universe will instantly emerge but with a degree of uncertainty equivalent to the amount entropy increased within the universe. This tells us that any predictable hypothesis cannot be precisely determined but approximated.

An increase in entropy from the big bang explosion to the present moment (t = 0) is equivalent to an amount of energy that had been degraded. But the present moment of certainty shows that it is the remaining undegraded energy that will be naturally degraded with time. For this we anticipate that rate of energy degrading after the present moment (t = 0) will be somewhat slower with time, as can be visualized by an illustrated diagram shown in Figure 6.7. It is because our universe expands at the speed of light, the net volume gain appears to slow down somewhat as the base volume V_0 enlarges, since $\alpha(t) = \Delta v/v_0$.

As entropy increases with time, energy within our universe will eventually be totally degraded, as our universe expands to the point of infinity. In other words, entropy within our universe will be continuingly increasing but at a slower rate. Nevertheless, entropy will reach infinitely large (i.e., $S \to \infty$) at $t \to \infty$, when there is no undegraded energy left behind to degrade. This means that all the useful undegraded energy that created our universe will be eventually degraded or used up. In other words, those degraded energy (or trash) will be eventually dissipated within the temporal (t > 0) space that our universe was embedded in. And this will be the end of our universe, from an entropy standpoint.

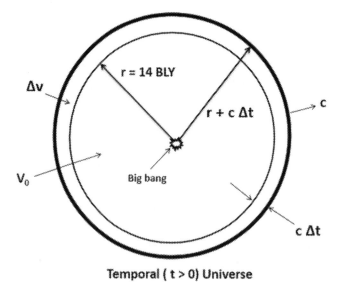

Temporal (t > 0) Universe

FIGURE 6.7   Shows that as our universe is expanding at the speed of light. But over time, net gain in volume Δv with respect to a larger base v_0 of our universal volume tends to slow down. This shows that the rate of entropy increasing will continuously taper toward zero at the point of infinity (i.e., $t \to 0$).

Since the present moment (t = 0) is the absolute certainty of our universe, this moment moves instantly (i.e., $\Delta t \rightarrow 0$) to the next moment of t = 0 + Δt but with an amount of entropy (i.e., ΔS) is added into it. And this amount of entropy increased represents a small degree of uncertainty as anticipated. In other words, the farther away from current certainty, the more uncertainty the prediction will be. Precisely this is what entropy increase means, prediction cannot be deterministic. Yet, without all those deterministic laws, principles, and theories, it seems to me there is no other better way to predict.

6.7 VIRTUAL PAST CERTAINTY

Nevertheless, entropy at the present moment (i.e., t = 0) is a lowest measure of absolute certainty moment (t = 0) [i.e., S(t = 0) = 0], but the complexity of the current universe is the highest. Even though the complexity had changed from a singular certainty to a very complex universe. The amount of entropy increased only described the overall complexity of our universe, but it does not represent the precise detail of our universe. Nonetheless, entropy is a necessary price to pay, such as Boltzmann's units or in Shannon's information measure [4,5,7]. But it does not give us the precise information of our universe.

The reason entropy increases with time naturally is our universe expands with time at the speed of light. But an amount of entropy increased within our universe is the equivalent amount of energy degradation within our universe. From this we see that complexity of our universe increases with time, yet the present moment is the lowest [i.e., S(t = 0) = 0]. However, it is the entropy complexity measure that has been used in the past to the present moment (i.e., t = 0), but beyond the present moment it is the remaining undegraded energy that increases the entropy. For example, if our universe had exhausted all the undegraded energy, then there is no way for our universe to increase the entropy. For this we see that it is the stochastic dynamic of our universe that changes the science, but not science that changes the dynamic of our universe. In words, without the stochastic dynamic of our universe, there will be no entropy increase and no science.

For simplicity, let me epitomize the nature of entropy, as depicted in Figure 6.8. In this figure, we see that the certainty moment (i.e., t = −14 BLY) started at the big bang hypothesis where entropy was at the lowest [i.e., S(t = −14 BLY) = 0] and our universe was expanding with time to the present moment (i.e., t = 0). As for the past 14 BLY I had dramatized all those expanded universes actually physically existed at various specific locations. The cone of ambiguity is to show all the possible predictable scenarios that could occur, yet the natural exclusive principle allows only one at a time to be physically realized. Since those are the physical facts of consequences, it is hard to accept the paradox of Schrödinger's cat [8]. It is trivial to see that a physical universe can only occur once at different moments of certainty, but cannot simultaneously coexist at the same certain moment. For example, can you simultaneously coexist with yourself yesterday, today, or even with yourself tomorrow although tomorrow has not arrived yet. And it is even more difficult to agree with Einstein's special theory [9], how can we move ahead or behind the pace of our universe since we are a part of the universe?

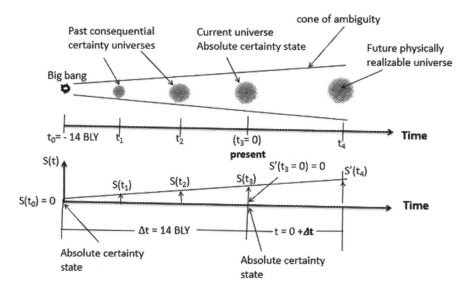

FIGURE 6.8 Epitomizes the stochastic dynamic of our universe. Upper layer shows entropy within our universe increases with time as projected by a cone of ambiguity. Within the cone shows the past certainty virtual universes that enlarge with time. Second layer illustrates the amount of entropy increases with time as our universe expands with time.

Nevertheless, the present moment is only moment of physical certainty, entropy is again the lowest, but it does not mean the complexity measures in entropy at present moment is the lowest. Since complexity of the current moment in entropy of our universe is the highest, it is not mean that the complexity measure provides the precise information of our universe at the present moment (t = 0). In this we see that entropy is a necessary cost in Boltzmann's units but not in real physical terms, since every subspace within our universe takes an amount of energy ΔE and a section of time Δt (i.e., Δt, ΔE) to create. Nonetheless, present moment certainty (t = 0) tells us that our universe still has a sufficient amount of undegraded energy to be degraded with time. In other words, when entropy reaches its maximum, there is no undegraded energy left for it to increase, as similar to a dead battery. From this we see that our universe is an irreversible subspace in time, as predicted by the second law of thermodynamics [10].

Nonetheless, entropy is an uncertainty principle, and once our universe had emerged with time to a physically real universe, our universe is the only absolute certainty that occurs at that time and so on. And this is precisely what temporal (t > 0) space is; it means that physical certainty continuingly goes on with the pace of time. From this we see that entropy at current physical certainty at t = 0 is the lowest [i.e., S(t = 0) = 0], since our universe is physically real at the present moment. This means that our universe is uncertain ahead on its next moves, as long as the remaining undegraded energy within our universe is not exhausted yet. Although the rate of entropy increase is somewhat slower as compared with the preview physical transformation. In other words, it is our universe that changes with time, but not our universe that changes time as in contrast with Einstein's general theory had stated that the time-space can be curved by space [9].

Nevertheless, entropy is a measure of complexity within an isolated subspace where time is momentarily stopped, but it is also a measure of uncertainty within the same isolated subspace where time moves on. From this we see that the past and future are either virtual or physically realizable but the present moment of absolute certainty (i.e., t = 0) divides them. And this is the amount of entropy increased (i.e., the complexity) the most since the big bang creation. Yet, it is the present moment that provides us with the lowest entropy to predict the future but uncertainty. In other words, without the increase of entropy, there will be no future hypothesis, and that must be the end of our universe.

So far science has developed from an empty subspace background inadvertently, for which we had buried ourselves within a timeless (t = 0) space for centuries. It is one of the objectives of this chapter, which presents to you the essence of temporal (t > 0) universe, as again illustrated in Figure 6.9. In this we see that our universe was created by a big explosion within a temporal (t > 0) space where the clock ticks. Our universe emerged from a small size that physical reality gradually changes when the clock ticks with a 14 BLY journey of each physically realizable changes until the current moment of certainty. From this it should be good enough to convince you that our universe expands with the clock ticks. And it is not our universe that changes the ticks.

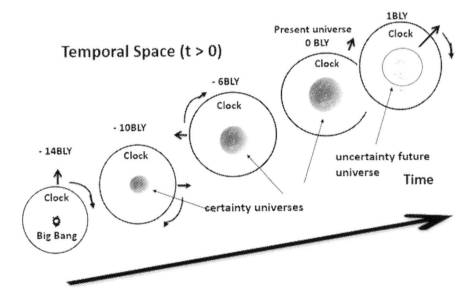

FIGURE 6.9 Illustrates one and only one absolute certainty universe continually expanding with the pace of time as the clock ticks. In this it shows that it is neither possible for two nor more certain universes to simultaneously exist with absolute certainty nor any of those absolute certainty universes can move ahead or behind. In this it also shows that a future physical realizable universe cannot be deterministic, yet general theory tells us it could.

In other words, our universe only physically existed of absolute certainty with time, where the pace of time is the only variable within our temporal space that cannot be changed. From which we see that, physical reality occurs only at current time of certainty. For example, physical universes were existed once at every precise past time of certainty shown in the figure. But future (t > 0) [i.e., (t = 0 + Δt)] universe will be physically realized yet with degree of uncertainty. In other words, future is unpredictable. For which we see that, the farther away from currently certainty the more uncertain to predict.

This is similar as we are moving within a train of time, either you are within or you are not a part of the time. For instance, if you are not a part of time, then you cannot exist within the physical reality of time or within our universe. From this we see that as we traced back to the past, we see that our universes certainly physically existed in a past time of certainty. And this is precisely why we used past certainty to look for a future hypothesis. But in view of our universe beyond the present moment of absolute certainty, its future may be predicted from the past certainty but it is not supposed to be exact or deterministic, since entropy within a universe increases naturally with time. In other words, our future universe is not a time-independent subspace, since it changes with time. And this is the part that we had missed for centuries, for which all the laws, principles, and theories are timeless (t = 0) or time independent. Although those laws and principles were the corner-stones of our science, at the same time they had created scores of fictitious and virtual sciences that do not exist within our universe. The fact is that our science had been hijacked by mathematics for centuries. But it is wrong to let mathematics lead our science, and instead we should have let physically realizable principles lead our hypotheses. Otherwise, we will forever be trapped within a timeless (t = 0) wonderland of fancy science, fictitious as mathematic is. For example, great scientists had even fantasized time traveling without a price to pay [11]. Or even send a signal without a section of time [12,13].

Nevertheless, the energy degrading probability with respect to the current absolute certainty is given by:

$$p(\Delta t')|_{\Delta t'=14 \text{ BLY} \leq 28 \text{ BLY}} = [1-\alpha(t')] \tag{6.26}$$

where I still used the linearized degradation curve for simplicity and t' denotes time after t = 0. For this we have a stochastic probability of p(Δt' = 0) as given by:

$$p(t = 0)|_{\Delta t'=0} = 1 \tag{6.27}$$

which shows that entropy at the present moment t = 0 is at the lowest state such as:

$$S(\Delta t' = 0) = 0 \tag{6.28}$$

As reference to the linearized curve, $\alpha(t)$ can be determined with the expansion of our universe, as illustrated in Figure 6.7, where we see a dynamic change in volume after the present certainty $t = 0$. For this, the rate of energy degradation within our universe based on certainty moment ($t = 0$) is given by:

$$\alpha(\Delta t) \approx \Delta v/v_0 \qquad (6.29)$$

where Δv is an incremental increase in volume and v_0 is the current volume of our universal at $t = 0$, as shown by:

$$\Delta v = (4/3)\pi[14 \ BLY + \Delta t]^3 - (4/3)\pi[14 \ BLY]^3 \qquad (6.30)$$

where 14 BLY is the time left after the present moment $t = 0$ as referred to the linearized curve shown in the figure. Since universal volume at present certainty (i.e., at $t = 0$) is given by:

$$v_0 = (4/3)\pi[14 \ BLY]^3 \qquad (6.31)$$

we have:

$$\alpha'(\Delta t') = \Delta v/v_0 = 3\Delta t'/14 \ BLY + 3(\Delta t')^2/(14 \ BLY)^2 + (\Delta t')^3/(14 \ BLY)^3$$

which is the rate of energy degrading of $E(t)$ after the present moment of certainty $t = 0$. Then the stochastic probability that represents energy degrading after $t = 0$ can be written as:

$$p(\Delta t')|_{\Delta t'=0 \leq 14 \ BLY} = [1-\alpha'(\Delta t')] \qquad (6.32)$$

Which corresponds to an amount of entropy increase:

$$S(t) = -k \ \ln \ p(\Delta t') \qquad (6.33)$$

In view of the relative rate of entropy increasing is dependent upon the ratio of $\Delta v/v_0$, we see that if the section of time Δt that is relatively smaller than 14 BLY as given by:

$$\Delta t' < < 14 \ BLY \qquad (6.34)$$

Thus, we have:

$$p(\Delta t') \approx 1-3\Delta t'/14 \ BLY \rightarrow 1 \qquad (6.35)$$

since $3\Delta t'/14BLY \approx 0$. From this we see that overall entropy increased within our universe is relatively small. Then a hypothetical prediction based on the current state of certainty would be closer to a deterministic solution. This was precisely what Einstein's general theory had done to predict planet Mercury's orbital path with a high degree of accuracy [14]. Although using deterministic theory to hypothesize is not a physically reliable approach, it had led us to believe that the general theory can change time-space. As in contrast within our temporal (t > 0) universe where space cannot curve time, but space changes with time.

If our prediction is with a range of $\Delta t \approx 2$ BLY, using Einstein's general theory to predict planet Mercury's orbital path would be very uncertain. Since after a couple billions light years, we are not certain that our solar system will still exist. Since the amount of entropy increase is very large after several BLY, in which stochastic probability is shown by:

$$p(\Delta t') \approx 1-(3\Delta t'/14 \text{ BLY}) \approx 0.57 \qquad (6.36)$$

that represents a 57% accuracy. As we see that, if the prediction is beyond 3 BLY (i.e., $\Delta t > 3$ BLY), then the prediction would be approximately equaled to 0%. From this we see that, it would be a mistake to use Einstein's deterministic general theory as applied to wormhole time traveling that some scientists had fantasized [11].

Nevertheless, at the present moment of certainty (i.e., t = 0), our universe still has undegraded energy E_0' left for entropy to increase. And this is the amount of total undegraded energy left behind that will be eventually degraded is given by:

$$\Delta t' \ \Delta E' = E_0' \qquad (6.37)$$

where $\Delta t' = E_0' - E(t = t')$, E_0' is the remaining undegraded energy at (t = 0), as referred to Figure 6.6. From this we see that the additional entropy increased is the part to be used for the uncertain entropy increased after t = 0. But it is not the remaining undegraded energy to overall contribute to complexity increase of our universe after the present moment, since overall energy is conserved:

$$(\Delta t = 28 \text{ BLY}) \ (\Delta E \approx ?) = (\tfrac{1}{2})Mc^2$$

which is the total energy conservation within our universe, where ΔE is a smaller amount of energy. From this we see that every hypothetical principle, theory, and law cannot be precise but approximated. Yet, as reference to energy conservation within our universe, there is a question remains to be asked: where are those degraded and undegraded energies located within our universe? Yet my assertion is that dark matter has a profound relationship with the degraded energy since our universe is an energy conservative subspace. While dark energy is related to the useful or undegraded energy and the rest must be the created observable substances (e.g., galaxies gases, particles, permeability permittivity, and others) [15,16].

6.8   ENTROPY STOCHASTIC DYNAMICS

Entropy is a measure of complexity and as well of uncertainty, but it cannot be both. Our temporal (t > 0) universe, from creation to beyond current absolute certainty is one of entropy stochastic dynamics examples. In terms of complexity, entropy is a spatial complexity measure where temporal (t > 0) physical certainty is not in consideration, although entropy increases with time. But in terms of uncertainty measure, entropy increases with time after a physical certainty has occurred, which is based on the remaining undegraded energy within the universe. And this is precisely what Boltzmann's entropy had stated: entropy increases with time within an isolated subspace.

In view of stochastic dynamics of our universe, as depicted in Figures 6.8 or 6.9, we see that our universe was created from a small but "certain" big bang explosion hypothesis and gradually expanded with time within a cone of ambiguity, as illustrated in the figure. Since time is a dependent variable with our expanding universe, entropy increases naturally with time. From this we see that our universe physically emerges with time but only one of the most probable physical realizable certainty. From this we see that the cone of ambiguity or uncertainty as depicted in the figure is based on equiprobable assumption, where entropy increases with time is the highest within our universe at that time. Yet, complexity does not actually represent the real physical stochastic dynamic scenario, since physical reality changes the probabilistic prediction to certainty. In other words, the next instantaneous physical transformation of our universe is not equiprobable, but based on the likelihood natural selection of the preceding certainty universe. In this we see that hypothesis cannot be absolutely certain, since certainty only exists at the moment of an actual physical reality appearance.

Nevertheless, natural instantaneously physical reality occurs once and only once at the most probable consequences. In other words, nature is not allowed to have more than two or more physical realities of our universe to exist at the same time, as in contrast with Schrödinger's fundamental principle. From this we see that our universe had never been virtual or uncertainty before, but this is by no means that we can go back to visit. Yet our universe's future cannot be absolute certain, no matter how small a section of time ahead of the present moment of time (i.e., $\Delta t \rightarrow 0$) for us to determine. This is precisely what physical reality means in our universe; this is the only existing moment of truth or at the present moment of absolute certainty. Similar to real-time tracking (i.e., $\Delta t = 0$), but real-time tracking is an absolute unrealizable science to attain. Yet, this is a dream that many quantum physicists would like to attain. But it is like chasing a beautiful timeless (t = 0) angle within our temporal (t > 0) universe, since timeless and temporal cannot coexist.

Thus, we see that entropy measures from the spatial standpoint it represents the complexity of our universe (or a subspace) by virtue of energy conservation. But it is nothing to do with the temporal (t > 0) certainty, since our universe instantly updates its physical certainty with time. From this we see that a physically realizable certain universe prevails regardless of degrees of uncertainty prediction. In other words, one of the most probable scenarios is emerging instantly (i.e., $\Delta t \rightarrow 0$) to absolute certainty, yet with degree statical errors.

Nevertheless, entropy increases after the present moment of certainty is based on the undegraded energy left behind within our universe. In other words, it is the total amount of undegraded energy left behind, but with a new constraint on energy conservation; for example, as shown by the energy degradation curve in

Figure 6.6(i.e., $\Delta t' \, \Delta E' = E_0'$). But the entropy increased is not to be added to the complicity entropy measure of our universe, since the complexity entropy is based on total spatial complexity of our universe at that time. From this we see there is a distinction between temporal (t > 0) and spatial entropy measure: one is entropy increase representing the complexity measure of our universe, and the other entropy increase is needed for prediction beyond the moment of certainty, as time moves on. Nonetheless, it shows that, uncertainty changes naturally with time, as in contrast with Heisenberg uncertainty [3] is based on observation or independent with time.

6.9 ENERGY CONSERVATION ENTROPY

Similarly, entropy increases with an isolated particle (or mass) in motion within our universe can be shown from the momentum conservation perspective is depicted in Figure 6.10. With reference to the dynamic probability of mass in motion which can be written by:

$$p(\Delta t) = E(t - \Delta t)/E_0 = [1 - (\alpha \Delta t/E_0)]$$

where $E_0 = (\frac{1}{2}) \, m \, v^2$ is the kinetic energy of m conserves within an isolated subspace that the particle is embedded in, and α is an arbitrary energy degrading rate which is assumed to be known. And $\alpha \Delta t / E_0$ represents a relative amount of kinetic energy degradation within a larger uncertain subspace after a section of time Δt.

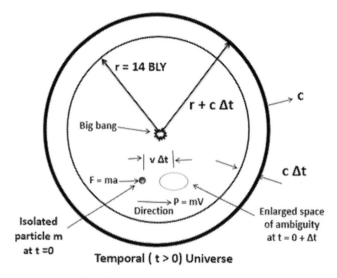

FIGURE 6.10 Shows a particle in motion within our temporal (t > 0) universe. p is momentum, m is mass, and v is the velocity. In this we see that ambiguity or uncertain increase as section of time Δt enlarges. This is influenced by the expansion of our temporal (t > 0) universe.

But every momentum decreases naturally with time within our temporal (t > 0) universe, there is an entropy increased due to kinetic energy degradation. We see that a larger uncertain subspace of the future prediction is depicted within the figure. In this we see that it has created a small temporal (t > 0) subspace (i.e., the sphere of ambiguity) within our universe. That created a temporal (t > 0) subspace within a larger temporal (t > 0) space which is our universe.

Since rate of entropy increases with time is dependent upon the total kinetic energy E_0 available within the subspace, we see that for smaller mass m, entropy tends to increase faster, since the rate of energy degradation $\alpha(t)$ is inversely proportional to the total energy $E_0 = (\frac{1}{2})mv^2$ that is:

$$\alpha(t) = E(t = \Delta t)]/E_0 \tag{6.38}$$

where E(t) is a rapid degrading energy function and we had assumed m is mass invariant. From this we see that a particle has a lower momentum and it has a faster the rate of energy to degrade. In other words, entropy increases faster for a smaller mass in motion. It is apparent that a smaller particle has a shorter momentum lifetime (i.e., smaller Δt), for which it is more difficult to be found.

As we accepted our universe changes with time, every isolated substance drifts naturally with time. For this we assume an isolated particle m is situated within our universe, as depicted in Figure 6.11. As time moves on from the present certainty moment to t = 0 + Δt, particle m moves naturally with time within an enlarged sphere of ambiguity or uncertainty. For this we see that Heisenberg's uncertainty is

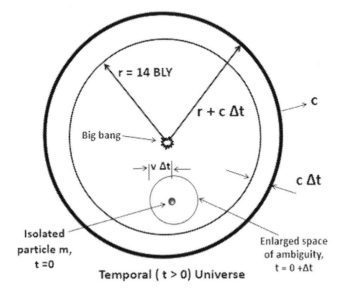

FIGURE 6.11 Shows a motionless particle m within our temporal (t > 0) universe. Since our universe expands with time, particle position drifts constantly with time within an ambiguity or uncertain sphere. This figure dramatizes that as section time Δt increases the uncertainty sphere enlarges.

an observational uncertainty principle [3]. But Boltzmann's entropy theory is a temporal (t > 0) uncertainty principle that changes naturally with time. From this we speculate again that the smaller the particle is, the more difficult it is to find since it drifts faster than a heavier particle with time.

Within a quantum world wavelets (e.g., photons) travel at the speed of light. From this every quantum leap starts by an isolated wavelet energy $E_0 = h\nu$ (i.e., a low entropy state), as illustrated in Figure 6.11. Since energy degradation increases as the ambiguous subspace enlarges, we see that an enlarging uncertain sphere expands with time as the speed of light is depicted in the figure. By virtue of energy conservation, within the uncertainty subspace, the bandwidth of quantum wavelet (i.e., $\Delta\nu$) decreases naturally with time. From this we see that $\Delta t \, \Delta E = h\nu$ or $\Delta t \, \Delta\nu = 1$ is conserved within the isolated sub-sphere. Since entropy increases more rapidly within the quantum regime at speed of light, it is more unpredictable within the quantum world. In other words, it has a smaller quantum leap energy (i.e., $h\nu$) to degrade Figure 6.12.

Similarly, an isolated directional quantum wavelet is depicted in the Figure 6.13, in which we see that the location of a short wavelet is within an enlarger uncertainty subspace that was due to an entropy increase. Since quantum state energy $E = h\nu$ travels at the speed of light, entropy of the traveling wavelet increases rapidly as the sphere of ambiguity enlarges with the pace of time. From this we see that it is very unlikely that multi-quantum wavelets can simultaneously exist at the same moment of time within our temporal (t > 0) universe. Once again, we see that Schrödinger's superposition principle cannot exist within a space that changes at the speed of light

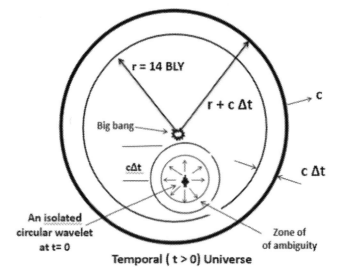

FIGURE 6.12 Shows a circular quantum wavelet travels at the speed of light within our temporal (t > 0) universe. We see that zone of ambiguity enlarged with time. But total energy $\Delta t \, \Delta E = h\nu$ is conserved within the zone of uncertainty. c is the speed of light and h is Planck's constant.

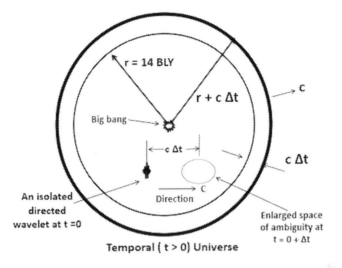

FIGURE 6.13 Shows a directional quantum wavelet travel at the speed of light within our temporal (t > 0) universe. An elongated sphere of uncertainty is shown at t = 0 + Δt. Within the uncertain sphere, quantum leap energy Δt ΔE = hv is conserved. c is the velocity of light.

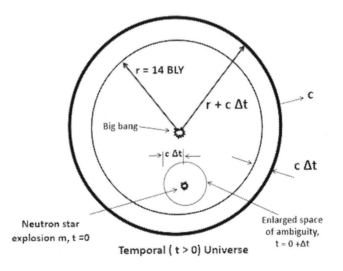

FIGURE 6.14 Shows a possible neutron star annihilation within our temporal (t > 0) universe. This is a much smaller-scale scenario but very similar to the creation of our universe. In this we see that all the exploded energy will eventually be degraded within the expanding subspace and dissipated within our universe, in view of conservation of energy.

with the same pace of time, since his principle is a timeless (t = 0) principle that has no time for entropy to increase [17,18].

Similar to the creation of our universe, we assume a neutron star had exploded within our universe [19,20], as depicted in Figure 6.14. From this we see that it creates a situation very similar to the creation of our universe, aside the electromagnetic waves expansion, it includes gravitational waves that travel at a much slower pace, and also other possible debris with it. As a whole, it is a bounded subspace that expands at the speed of light with possible detectable gravitation waves, similar to recently reported gravitational wave due to star annihilation [21]. Since the total energy of an annihilated neutron star is relatively very small compared with the giant galaxy, we had anticipated an entropy increase very rapidly and quickly dissipated within our observable sky. This is a very similar scenario to what our universe eventually will be, but it is a much smaller scale example.

One of the intriguing aspects in quantum theory must be the quantum entanglement, in which two quantum wavelets can be entangled regardless how far they are separated, which is essentially Pauli's exclusive principle [22]. But Pauli's principle was developed from an empty space platform where time does not exist, which is similar to all the laws and principles derived from it. For example, if we let a quantum leap energy radiate in the opposite direction within our temporal (t > 0) space, as illustrated in Figure 6.15, we see that that this set uncertainty wavelets may be detected in difference distances at d' = c $\Delta t'$ and d = c Δt, respectively. Because of limited bandwidth of quantum leap reality, it would be very unlikely that these wavelets travel at the speed of light can be detected at the same time, since within our universe entropy increases with time. From this we see that instantaneously and simultaneously entanglement can only exist within a timeless (t = 0) subspace, which has no time and no distance. This is precisely where Pauli's exclusive principle developed from. However, from the temporal exclusive principle, timeless subspace cannot exist within our temporal universe.

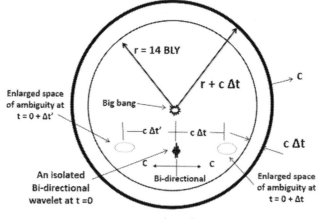

Temporal (t > 0) Universe

FIGURE 6.15 Shows a set of quantum wavelets created from a quantum leap, as directed in opposite direction, which travel at the speed of light within our temporal (t > 0) universe. The elongated spheres hypothesized the ambiguity of the quantum wavelets will be t = 0 + $\Delta t'$ and t = 0 + Δt. Within each uncertain sphere, quantum leap energy Δt $\Delta E = hv$ is conserved. c is the velocity of light.

In view of the examples shown in Figures 6.11 to 6.15, we see that essentially we had created isolated temporal (t > 0) subspaces within our temporal (t > 0) universe. It tells us that if any created subspace is not temporal (t > 0) subspace, they cannot exist within our temporal (t > 0) universe. Secondly, we had seen those uncertainty subspaces are profoundly connected to increasing entropy within our universe. In other words, without the continual expansion of our universe, the entropy increase within our universe would have to stop, since entropy increased is depending on energy degradation. Nevertheless, before exhausted all the undegraded energy within our universe, may be a new universe might have already started somewhere in the vast temporal space that our universe is embedded in. And this could have been happening even before the disappearance of our universe at point of infinity (t $\rightarrow$ 0). Since within the vast temporal (t > 0) space it should have a huge amount of degraded energy available can be undegraded. From this I hypostasize, given a right condition, those degraded energy can be reversed back to mass, since energy and mass are equivalent. Nevertheless, it is justifiable to say that all the degraded energies as well with other available substances, in principle, can be converted back into mass. In other words, a created gigantic mass will eventually be exploded to create a brand-new universe, somewhere within the vast temporal space, and so on.

6.10 A HYPOTHETICAL CREATION

Since entropy increased within an isolated subspace is a natural thing due to the expansion of our universe, all the degraded and leftover undegraded energy within our forever expanding universe will eventually dissipated within a larger temporal (t > 0) space that our universe had embedded in. By definition of the second law of thermodynamics [10], our universe will eventually seized to exist. Although our universe had exhausted most its undegraded energy (i.e., useful energy) that created our universe, but from an energy conservation standpoint we see that all those degraded energy has to be returning back to the greater temporal (t > 0) space that supported our universe. From this we see that there must scores of degraded energies that can be found currently within our universe, even well before our universe totally dissipated within the greater temporal space. Since all those degraded energies are in the form of substances, this could eventually be converted back to useful (i.e., undegraded) energy. For this we see that some of those degraded energies and possibly other substances or debris can eventually be finding their ways to create a new concentrated mass, for example such as black hole [22] and others. Eventually, over time, a gigantic mass could have created even before our universe has exhausted its undegraded energy. In other words, it is possible those degraded energies and possibly other leftover debris from a dying universe will eventually convert into mass, since the greater space that our universe embedded in is not timeless (t = 0) but an alive temporal (t > 0) space. Otherwise, the life cycle of universes will end forever, which is very unlikely.

Since energy and mass are equivalent, we see that converting degraded energy into mass can be written as given by [1,2]:

$$\frac{\partial m}{\partial t} = \frac{2}{c^2}\frac{\partial E}{\partial t} = -\frac{2}{c^2}[\nabla \cdot S(v)] \approx -\frac{2}{c^2}[\nabla \cdot (\mathbf{E} \times \mathbf{H})] \tag{6.39}$$

where c is the velocity of light, m is mass, E is the degraded energy, v is the frequency, $(-\nabla)$ is a convergent operator, S (v) is a degraded-energy vector, ($\cdot$) is the dot product, $\mathbf{x}$ is the cross product, and $(\mathbf{E} \times \mathbf{H})$ is the Poynting vector [23]:

$$(\mathbf{E} \times \mathbf{H}) = -\frac{2}{c^2}\frac{\partial}{\partial t}\left[\frac{1}{2}\varepsilon E^2(v) + \frac{1}{2}\mu H^2(v)\right] \tag{6.40}$$

$[E (v), H (v)]$ are the electromagnetic (EM) waves, and (ε, μ) are the permittivity and permeability within the greater temporal space that our universe once had embedded in.

With reference to degraded energy to mass conversion of Eq. (6.39), it shows that an intense electromagnetic energy is converging into a small volume to form a quantity of mass m. We see that it will take a huge amount of degraded energy to eventually converted into gigantic mass for another big bang explosion. This is a very slow process, since energy to mass conversion is inversely proportional to c^2 that takes a long time to make happen. But within the vast cosmological space, availability of huge amounts of degraded energy and other debris had never been a shortage.

Yet, it is must be the intense gravitational field induced by the converted mass to eventually ignite another big bang explosion. This hypothesis is exactly shown in preceding chapter that how our universe was created by the big bang hypothesis. We had shown without the indued gravitation field it would be very difficult to justify that the big bang explosion had ignited by itself. Which is in contrast with commonly believes that the big bang creation was started from an empty space paradigm. But firstly, empty space has no substance for gravitational field to exist. Secondly, empty space is not a physically realizable paradigm.

Nonetheless, everything within our universe no matter how small it has a life. From which we see that every particle will eventually decay or disintegrated within our universe. In other words, particle's energy (i.e., ΔE) will eventually be degraded within our universe, but not disappear by means of energy conservation.

6.11   TOPIC TO INVESTIGATE

The purpose of this book is to let theoreticians know that it is about time for us to look into the physically realizable science, otherwise we will be forever trapped within an unsupported timeless (t = 0) fantasy science, although those deterministic principles and laws are still the cornerstones of our foundation. Since scores of virtual and fictious principles and theories are available in the science community, allow me to single out one of the most trivial consequences that was developed from a time-dependent subspace. This could be an exciting starting point, otherwise you may not be interested. This must be the quantum electrodynamics (QED) developed by the legendary Richard Feynman [24].

For this, let us start with the Feynman diagram, as depicted in Figure 6.16, where we see an electron and a positron eventually annihilated produce a photon that created a quark and an anti-quark particle and so on. This must be an exciting discovery if the Feynman diagram is embedded within a temporal (t > 0) subspace.

Since Feynman's equations were developed based on a non-physically realizable paradigm, I am in doubt that his equations are legitimated within a temporal (t > 0) space, in spite of the complexity of mathematics. For example, when an electron and positron were annihilated, my question is that, where does the energy go, since electrons and positrons are matter. In view of energy is equivalent to mass, a huge amount of energy should explode. Added is the embedded substances are not changed naturally with time, and it is not a physically realizable paradigm. And this is the reason why all Feynman QED equations are deterministic, yet prediction is supposed be approximated. Particularly within the quantum regime, things change at the speed of light with the same pace of time.

The proposed scenario can be started by submerging the Feynman diagram within a temporal (t > 0) space, as depicted in Figure 6.17, which is basically a physically realizable paradigm. But several questions remain to be fixed. Firstly, antiparticles travel backward in time, we see that matter and anti-matter can only coexist at t = 0, but timeless (t = 0) space is a virtual mathematical space. Secondly, when electron and positron annihilated where all those energies go, since electron and positron are matters. Since every subspace entropy increases naturally, then Feynman's equations are not supposed to be deterministic. Particularly within the quantum electrodynamic regime, where particles change at the speed of light. For these reasons I think it is an interesting topic for a mathematically inclined scientist to explore. I am every certain that they will eventually come up with a set of different equations, which are more justifiable to exist within our temporal (t > 0) universe. In that sense, some of the so-called jewels of physics as derived from the QED equations may not even exist or need to re-configure, since Feynman's equations were developed from a time-independent subspace. And good luck.

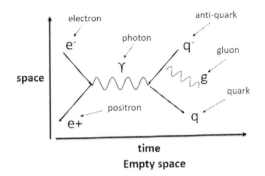

FIGURE 6.16 Shows a Feynman diagram embedded within a time-independent space paradigm. As from the temporal (t > 0) exclusive principle, it is not a physically realizable paradigm, since a particle cannot coexist with time-independent space.

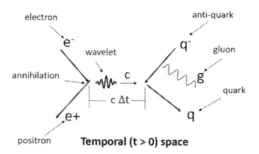

FIGURE 6.17 Shows Feynman diagram is embedded within a temporal (t > 0) space. Basically, this is a physically realizable paradigm that should be used for Feynman analyses, since substance in consideration can coexist within our temporal (t > 0) universe. But questions remain, as those substances obey the temporal (t > 0) condition, such as changes with time, momentum, and energy conservation criteria, and possibly others.

6.12 REMARKS

I would summarize that Boltzmann's entropy is not a mystery after all as it stated. Every principle, theory, and law changes naturally with time, and Boltzmann's entropy cannot be the exception. Uncertainty is a natural thing that cannot be stopped, since our universe is a temporal (t > 0) stochastic dynamic subspace that expands at the speed of light within a greater temporal (t > 0) space. Since our universe is a bounded isolated expanding space, it affects every subspace within the universe. For this entropy within every isolated subspace increases naturally with time. But energy within an isolated space degrades with time, and we see that an amount of energy degraded is equivalent to an amount of entropy increased. For example, as our universe expands at the speed of light, an amount of energy had been degraded, which is equivalent to an amount of entropy increased. From this we see that Boltzmann's entropy told us that space and time coexisted, as in contrast with most of scientists who are still believing that time is an independent dimension.

Entropy increases within our universe and cannot be stopped, except at the point of infinity (i.e., t → ∞). Yet, conversion from degraded energy back to mass may have started well before the exhaustion of all the undegraded (useful) energy. From this we see that over time it is very likely that a new universe created somewhere within the greater temporal (t > 0) space has been started. There is no other way to hypothesize future consequences without using past certainty (i.e., information), but deterministic principles produce exact solutions. The irony is that all the laws and theories are exact. These must be the reason scores of their solutions were virtual as mathematics. Yet, it seems to me that we have lost our logical, independent thinking and have opted to accept the approval of the others. Maybe it is about time for us to look into the physically realizable science; otherwise we will be forever trapped within the timeless (t = 0) wonderland of fantasy science, where entropy does not apply since it has no section of time Δt and an amount of energy ΔE to pay.

REFERENCES

1. F. T. S. Yu, "Time: The Enigma of Space", *Asian Journal of Physics*, 26 (3): 143–158, 2017.
2. F. T. S. Yu, "From Relativity to Discovery of Temporal (t > 0) Universe", *Origin of Temporal (t > 0) Universe: Correcting with Relativity, Entropy, Communication and Quantum Mechanics*, Chapter 1, CRC Press, New York, 1–26, 2019. New York.
3. W. Heisenberg, "Über den anschaulichen Inhalt der quantentheoretischen Kinematik und Mechanik", *Zeitschrift Für Physik*, 43 (3–4): 172, 1927.
4. L. Boltzmann, "Über die Mechanische Bedeutung des Zweiten Hauptsatzes der Wärmetheorie", *Wiener Berichte*, 53: 195–220, 1866.
5. C. E. Shannon and W. Weaver, *The Mathematical Theory of Communication*, University of Illinois Press, Urbana, IL, 1949.
6. F. T. S. Yu, *Optics and Information Theory*, Wiley -Interscience, New York, 1976. *Information Theory*, Wiley
7. L. Brillouin, *Science and Information Theory*, 2nd edition, Academic Press, New York, 1962.
8. E. Schrödinger, "An Undulatory Theory of the Mechanics of Atoms and Molecules", *Physical Review*, 28 (6): 1049, 1926.
9. A. Einstein, *Relativity, the Special and General Theory*, Crown Publishers, New York, 1961.
10. F. W. Sears, *Thermodynamics, the Kinetic Theory of Gases, and Statistical Mechanics*, Addison-Wesley, Reading, Mass, 1962.
11. M. S. Morris and K. S. Thorne, "Wormholes in spacetime and their use for interstellar travel: A tool for teaching general relativity", *American Journal of Physics*, 56 (5): 395–412, 1988.
12. C. H. Bennett, "Quantum Information and Computation", *Physics Today*, 48 (10): 24–30, 1995.
13. K. Życzkowski, P. Horodecki, M. Horodecki, and R. Horodecki, "Dynamics of quantum entanglement", *Physical Review A*, 65, 1–10, 2001.
14. G. M. Clemence, "The Relativity Effect in Planetary Motions", *Reviews of Modern Physics*, 19 (4): 361–364, 1947.
15. G. Bertone, Editor, *Particle Dark Matter: Observation, Model and Search*, Cambridge University Press, Cambridge, UK, 2010.
16. L. Amendola and S. Tsujikawa, *Dark Energy: Theory and Observation*, Cambridge University Press, Cambridge, UK, 2010.
17. F. T. S. Yu, "Nature of Temporal (t > 0) Quantum Theory: II", *Quantum Mechanics*, Edited by P. Bracken, Chapter 9, pp. 161–188, IntechOpen, London, 2020.
18. 8. F. T. S. Yu, "The Fate of Schrodinger's Cat", *Asian Journal of Physics*, 28 (1): 63–70, 2019.
19. A. Heger, C. L. Fryer, E. E. Woosley, N. Langer, and D. H. Hartmann, (2003). "How Massive Single Stars End Their Life", *Astrophysical Journal*, 591 (1): 288–300, 2003.
20. B. P. Abbott, et al. (LIGO Scientific Collaboration & Virgo Collaboration), "GW170817: Observation of Gravitational Waves from a Binary Neutron Star Inspiral", *Physical Review Letters*, 119 (16): 2017.
21. M. Bartrusiok, *Black Hole*, Yale University Press, New Haven, CT, 2015.
22. W. Pauli, "Über den Zusammenhang des Abschlusses der Elektronengruppen im Atom mit der Komplexstruktur der Spektren", *Zeitschrift für Physik*, 31, 765, 1925.
23. J. D. Kraus, *Electro-Magnetics*, McGraw-Hill Book Company, New York, 1953, p. 370.
24. R. P. Feynman, *Quantum Electrodynamics*, Addison Wesley, Cambridge, Mass, 1962.

Appendix A: Essence of Temporal (t > 0) Exclusive Principle (TEP)

Since our universe was created from a huge mass to energy conversion, our universe is a stochastics dynamic temporal (t > 0) subspace that expands at speed of light. In other words, our universe is not a zero-sum energy empty space, but an energy conservation subspace where everything (i.e., substance) that exists within our universe is temporal (t > 0); otherwise, it will be excluded from our universe. This is the temporal exclusive principle that I advocated in this book, from which I had shown that laws, principles, and theories developed from an empty space; for example, as depicted in Figure A.1, will be timeless or time-independent. Since substance and empty space cannot coexist, we have seen that an empty space paradigm is not a physically realizable paradigm. And this is precisely the reason why all the principles and theories that we developed are either timeless or time-independent; strictly speaking there are not physically realizable laws and principles since we cannot implement them directly within our temporal (t > 0) universe. In fact, some of them cannot even exist within our universe, fictitious as mathematical logics; for example, Schrödinger's Cat and Einstein's time-dilation and curves time-space.

Nevertheless, if we submerge this particle in motion within our temporal (t > 0) universe as depicted in Figure A.2, we see that an induced magnetic field cannot be ignored. Secondly, velocity of the moving particle is changing with time, yet total energy is conserved as has been shown in Chapter 6, since it is an energy conservation subspace.

Similarly, if we submerge a charged particle (e.g., electron) in motion within our universe, as depicted in Figure A.3, we see that it has an induced electric field with an induced gravitational field associated with the moving electron, which cannot be established as embedded within an empty space.

Energy transmission within a quantum leap regime is at light speed; a physically realizable condition is more stringent, as can be seen in Figure A.4. From this, we see that it was wrong for submerging a quantum leap atomic model within an empty space paradigm, where E = hv is the total energy of a quantum leap, but it does not represent a physically realizable wave function since it is not temporal (t > 0).

Nevertheless, temporal (t > 0) universe is an energy conservation subspace, and it is not a zero-summed energy subspace. For which I see that, any anti-matter as speculated by theoretical physicists, cannot be existed within our universe.

Let me show a universe paradigm that has been used for centuries, as depicted in Figure A.5. Firstly, let me note that it is not a physically realizable paradigm, by virtue of the temporal exclusive principle. Normally, we assume deep space is

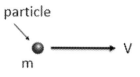

Classical Paradigm
(a piece of paper)
Empty space
Timeless (t =0) subspace

FIGURE A.1 Shows a particle in motion within an empty space, which is not a physically realizable paradigm in view of the temporal exclusive principle (TEP). m is mass and v is velocity of the particle.

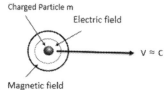

Physically realizable paradigm
Non-Empty space
Temporal (t >0) subspace

FIGURE A.2 Shows a particle in motion within a temporal (t > 0) space, which is a physically realizable paradigm. m is mass, v is velocity of the particle, and c is the speed of light.

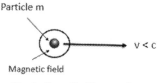

Quantum regime paradigm
Non-Empty space
Temporal (t >0) subspace

FIGURE A.3 Shows a charged particle in motion within a temporal (t > 0) space, which is a physically realizable paradigm. m is mass, v is velocity, and c is the speed of light.

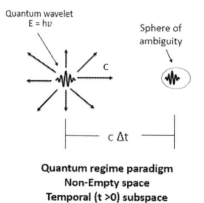

FIGURE A.4 Shows a quantum wavelet radiator situated within a temporal (t > 0) space, which is a physically realizable paradigm. Δt is a section of time and c is the speed of light, in which we see that wavelet changes with time.

Quantum regime paradigm
Non-Empty space
Temporal (t >0) subspace

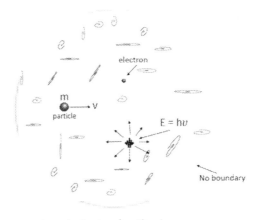

A static timeless (t = 0) universe

FIGURE A.5   Shows a particle in motion, an electron, and a quantum leap radiator situated within a static universe which has np boundary. By virtue of the TEP, it is not a physically realizable paradigm. m is mass and v is velocity of the particle.

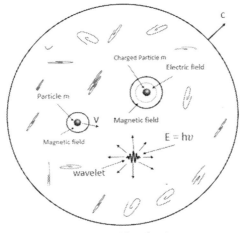

Our temporal (t > 0) universe

FIGURE A.6   Shows a particle in motion, an electron, and a quantum wavelet situated within our dynamic temporal (t > 0) universe, where the boundary expands at the speed of light c. By virtue of the TEP, it is not a physically realizable paradigm. m is mass and v is velocity of the particle.

empty, and it is impossible to create an induced gravitational field by mass as well as electric field by charge. From this we see that any hypothetical solution that comes from this paradigm very likely is not physically realizable.

Nevertheless, if physical substances are submerged within our dynamic temporal (t > 0) universe, as depicted in Figure A.6, we see that the induced gravitational and electric field, as well quantum wavelet, have to be included for the postulated analysis. Any solution that comes out from this paradigm is likely to be physically realizable.

From this we found that substances, regardless of their size, within our temporal (t > 0) universe behavior are basically the same. It is, however, the total dynamic energy that determines the rapid changes with time, as described in Chapter 6. For example, although global changes are relative slower for a heavier mass within the same section of time Δt, the complexity (e.g., particles) within the substance changes more rapidly. This is very similar to the dynamic changes of our universe with time, as shown in Chapter 5. For instance, a 1-second global change of our universe seems insignificant, yet within our universe it changed a lot within 1 second.

Appendix B: Aspects of Particle and Wave Dynamics

In view of a temporal (t > 0) universe, every subspace within our universe is a time-dependent subspace, in which time coexisted with every subspace and time is a "dependent forward variable". The speed of light within our universe is dictated by the pace of time that our universe was created with. In other words, any subspace (i.e., substance) within our universe is a time-dependent or temporal (t > 0) subspace in which time speed within every subspace is in unison with the pace of time within the entire universe; otherwise, the subspace cannot exist within our universe. For example, time speed within a subspace at the edge of our universe is at the same pace as the subspace closer to the center of our universe. On the other hand, if time speed within a subspace runs faster or slower than the time speed of our universe, then the subspace cannot be existed within our universe. An extreme example is that; a timeless (i.e., empty) space is a virtual mathematical space; firstly, it is not a temporally realizable space and secondly the space has no time for which it cannot exist within our temporal (t > 0) universe. For this, I named it the temporal (t > 0) exclusive principle (TEP principle). That is, temporal and timeless space cannot coexist.

Since every subspace within our universe was created by an amount of energy ΔE and a section of time Δt, it cannot bring back the section of time Δt that had been used for the creation. There is, however, a profound relationship between energy and mass (i.e., ΔE, Δt) by virtue of the mass-energy equivalent equation [i.e., $E = (1/2) mc^2$]. There exists a duality between energy ΔE and bandwidth Δυ. Nevertheless, without the coexistence with time, the duality collapses at a timeless (t = 0) empty space, which is not a subspace within our universe.

Let us look at energy ΔE and bandwidth Δυ duality; firstly it must be the existence of waves, otherwise it will not have the aspect of bandwidth. So what is a wave? A wave can only exist within a temporal (t > 0) subspace, since physical substance and time coexist; otherwise, waves cannot propagate with time. In other words, a wave can only exist within a temporal medium (i.e., substance) that creates a wave. This is also one of the key factors that many quantum physicists believe—that quantum leaps radiation propagates within an empty space. Aside from the non-physically realizable issues, substance (i.e., radiation) and empty space are mutually exclusive, the fact is that a wave cannot exist within an empty (i.e., timeless) space. Secondly, every physical radiation has to be band and time limited (i.e., Δυ, Δt,), as given by the Heisenberg Uncertainty Principle:

$$\Delta\upsilon \cdot \Delta t \geq 1$$

Yet, the Heisenberg principle is an observational principle, which is independent with time. But the physical significance of his equation had turned out the same as changes with time that I had explained in the preceding chapter. Nevertheless, we see that there exists an energy and bandwidth relationship as given by:

$$\Delta E = h\Delta \upsilon$$

where h is the Planck constant. This is the well known quantum leap energy or a "quanta" of electro-magnetic radiation, in which we see that a time-limited package of energy ΔE travels at the speed of light within our temporal $(t > 0)$ universe.

Nevertheless, the root of particle-wave duality was originated by the acoustic wave dynamics were acoustic wave travels "longitudinally". For example, an acoustic wave generated by a guitar is produced by a vibrating string that satisfies the wavelength selectivity boundary condition that causes a pulse of wavelet (i.e., time and band limited) that travels within our atmospheric space. Since it is a longitudinal wave, an acoustic wave depends on media such as solid material, liquid, or air for transmission. In other words, without such media, any acoustic wave cannot be created within the medium. Thus, we see that the particle-wave dynamics is a mathematical description by using the dynamics wave propagation to predict the behavior of an assumed particle agitation. Notice that our illustration is in fact a fix-ended string vibration instead of a particle, in which we see that it is actually a string vibration to wave dynamics.

On the other hand, an electromagnetic wave is a transversal wave that propagates within an electromagnetic (EM) medium (i.e., permeability ε and permittivity μ medium). Again, we note that empty space has no medium in it; an electromagnetic wave cannot exist within an empty space. In view of a particle-wave duality description, electromagneticwave dynamics exist within our temporal $(t > 0)$ space, and it is not the implication that an EM wave was created by means of a physical particle (i.e., photon) agitation within the space, since every physical particle requires a rest mass. Although the notion that a photon behaves as a particle has been well accepted, but it is difficult to reconcile with the relativity theory for the assumption; a photonic particle has an empty mass. For this, I would regard every photon as a virtual particle that is attached with a quanta energy $h\Delta\upsilon$. Even though mass and energy are equivalent from the mass-energy equivalent equation, a quantum of electromagnetic energy is not created by the annihilation of a rest mass; instead, it is in the form of EM energy releases by each quantum leap radiation $h\Delta\upsilon$. Which is a packet of energy ΔE to wavelet $\Delta\upsilon$ duality to describe the dynamics of quantum state behavior. Since energy has different forms such as potential, kinetic, chemical, radiation, nuclear and others, photonic energy is a packet of wavelet radiation derived from a quantum leap of an electron within an atom. In other words, particle-wave duality describes a packet of energy ΔE to wavelet (i.e., $\Delta\upsilon$) dynamics or simply $(\Delta E, \Delta\upsilon)$ duality. The energy transfer from a higher quantum state to a lower quantum state releases energy in EM wave, very similar to a fix-end string oscillation from a guitar. Therefore, it is more suitable to use energy for wavelet $(\Delta E, \Delta\upsilon)$ duality to describe the quantum leap wave dynamics; otherwise, the particle-wave duality gives us a notion that photon is a physical particle instead of a quantum of energy, in which we see that a time limited wavelet represents a package of energy, and a package of energy is a time-limited wavelet.

Furthermore, if the size of a pinhole is smaller than the wavelength of an illuminator, would you able to observe the diffraction pattern? If the answer is no, then we see that wave dynamics is equivalent to particles in motion but not equal to a particle since a photonic particle has no size. From this we see that a particle in motion is equivalent to wave dynamics, but a wave is not a particle and a particle is not a wave.

Nevertheless, theoretical physicists treated the quantum wave function as a means to speculate or predict its wave-particle duality. But firstly, a wave is equivalent to particle dynamics but a wave is not actually a physical particle in motion. Secondly, a wave or wavelet physical structure has a limited amount of information for us to predict its virtual particle dynamics with certainty, since every wave or wavelet is temporal ($t > 0$) and changes with time. For this it is wrong to predict photonic (i.e., virtual particles) distribution with certainty since distribution changes naturally with time within our temporal subspace.

Since particle-wave dynamics does not mean that a particle in motion is equal to a wave, or a wave is equal to a particle. This is similar to what mass annihilation to energy dynamics means, that energy and mass are equivalent, but it is by no means mass equals energy. And this precisely is what particle-wave dynamics means; it is a quantum leap energy from a fixed-ended mathematical representation of a wave. In other words, it is a time-limited package of wavelet energy $\Delta E \, \Delta t = h\nu$, but it is not intended to be equal to a particle in motion. Since electronics has been extremely successfull in applications, we fixated that quantum leap energy is electronic, and named it photonics. We obsessed that the quantum field regime would behave like electronics at the speed of light. This erroneous notion had been created as a score of fictitious ideas that photons are particles like substances that can event capture or store within an isolated subspace, instead of knowing that the package of quantum leap energy will be degraded, by virtue of energy conservation where entropy increases naturally. Maybe it is about time for us to redefine quanta as a package of quantum leap energy instead of photons, which sounds like an electron. Although $\Delta E \, \Delta t = h\nu$ is similar to the momentum conservation $\Delta E \, \Delta t = (\frac{1}{2})mv^2$, v is the velocity, or to mass-annihilation conservation $\Delta E \, \Delta t = (\frac{1}{2})mc^2$, c is the speed of light. Nonetheless, particle-waves dynamics are equivalent, but nor equal. For example, wave is an energy vector, while particle is a substance. In other words, particle can be isolated wave cannot. From which we see that, photon is equivalent to a particle in motion, but photon cannot to isolated, since it represents a package of wavelet energy.

Appendix C: What Temporal Space Does to Hypotheses

Firstly, let me note that, we discovered laws of science, but we did not create laws of science. For example, we discovered the temporal (t > 0) universe, but we did not create the temporal (t > 0) universe. From this we see that it is what the universe does to hypothetical theory, but the hypothetical theory does not change the laws of nature. In other words, science changes with the law of nature, but does not violate the law of nature, yet all laws we create are approximated, which includes the law of temporal exclusive principle. Nevertheless, the temporal exclusive principle (TEP) we discovered is closer to the truth of all the existent principles.

Since the Feynman equations [1] were developed from a time-independent subspace paradigm, as shown in (Figure B.1), it is not a physically realizable paradigm. It is very unlikely a solution obtained from this paradigm that would be physically realizable.

Nevertheless, it is a not physically realizable model; what annihilates the electron and the positron particles as they mated? To be more precise, does annihilation only involve electric field energies, masses, or both? Secondly, if we treated a positron as an anti-matter, then it cannot exist within our temporal (t > 0) universe, since our universe is an energy conservation subspace. Any anti-matter cannot exist within an energy conservation subspace, since anti-matter only exists within a zero-sum energy subspace, but our universe is not. This is also one of the many examples that shows the empty space paradigm produces a non-physically realizable solution. Since many astrophysists assumed our universe was created within an empty space paradigm, this means that our universe is a zero-sum energy subspace. But it is energy within our temporal (t > 0) universe that degrades naturally with time, for which energy cannot be annihilated within our universe, as shown in Chapter 6.

Now, if the Feynman equations have developed within a temporal (t > 0) subspace, this is depicted in Figure B.2. Generally speaking, it is an acceptable physically realizable paradigm, since temporal substances can exist within a temporal (t > 0) space, since electrons and positrons are physical particles [i.e., temporal (t > 0)].

Nevertheless, electrons and positrons are charged particles; firstly, they should have induced electric fields with them, as diagramed in Figure B.3. For this we see that the induced opposite electric field energies neutralize or annihilate to produce gamma ray radiation, as the Feynman diagram suggested. Since every radiator is limited by a section of time Δt and an amount of energy ΔE [i.e., Δt, $\Delta E(t)$], the emitted gamma ray changes naturally with time.

Nevertheless, electrons and positrons are particles. I presumed the annihilation must be due to electric field neutralization, which has nothing to do with mass. However, if it is due to mass-annihilation, it will produce a broad spectral band of radiation, but only the gamma ray is shown in the Feynman diagram. From this, as I

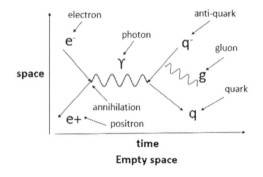

FIGURE B.1   Shows a Feynman diagram.

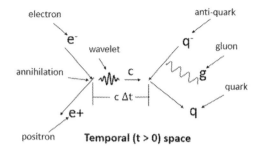

FIGURE B.2   Shows a physically realizable Feynman diagram.

see it, a gamma ray is the product from neutralizing the electrical field energies as electrons collided with positrons.

Since our universe is an energy conservation subspace, all the anti-matters, negative energy as proposed by theoretical physicists cannot exist within our temporal ($t > 0$) universe. As all the QED equations were developed from a non-physically realizable paradigm, I wonder if those hypothesized sub-atomic particles developed from this set equations actually existed within our universe. For example, anti-matter cannot exist within our temporal ($t > 0$) universe, by virtue of the temporal ($t > 0$) exclusive principle.

CHARGED PARTICLES IN TEMPORAL SPACE

Since our universe cannot be absolute empty, everything within our universe must be temporal ($t > 0$). In other words, every law and principle must comply with the temporal exclusive principle. This includes all the substances that are not in particle form, such as permeability $\mu(t)$, permittivity $\varepsilon(t)$, and possible others, since they are all created by an amount of energy ΔE and a section of time Δt. This means that ΔE (t) and $\Delta t(t)$ coexisted and are temporal ($t > 0$).

Regardless of the size, every charged particle induces electric and gravitational fields within our universe. For this, as a particle moves, its induced fields move with the particle. For example, as an electron spins or moves, induced electric and

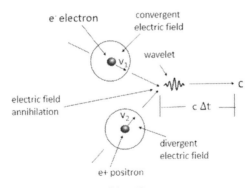

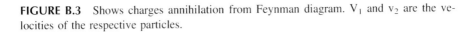

Temporal (t > 0) space

FIGURE B.3 Shows charges annihilation from Feynman diagram. V_1 and v_2 are the velocities of the respective particles.

magnetic fields move with the particle. This is one of the reasons that we have not found a magnetic single-pole particle, and maybe a magnetic single-pole particle cannot exist within our universe.

Similarly, Hawking's radiation [2] is a legacy of Einstein's general theory, which is not a physically realizable principle. It may be interesting enough for someone to show Hawking's hypothesis may not exist within our temporal (t > 0) universe, since the general theory was developed from an empty subspace platform.

In view of QED equations and derivations of Hawking's radiation, we see that it is not how rigorous mathematics is, it is the temporal (t > 0) subspace paradigm that determines their solutions are physically realizable. From this I stress that just depending upon mathematics alone cannot guarantee the solution would be physically realizable.

Nevertheless, electrons and positrons are charged particles; firstly, they should have induced electric fields with them, as diagramed in Figure B.3. For this we see that the induced opposite electric field energies neutralize or annihilate to produce gamma ray radiation, as the Feynman diagram suggested. Every radiator is limited by a section of time Δt and an amount of energy ΔE [i.e., Δt, $\Delta E(t)$], from which the emitted gamma ray changes naturally with time.

REFERENCES

1. R. P. Feynman, *Quantum Electrodynamics*, Addison Wesley, Cambridge, Mass, 1962.
2. S. W. Hawking, "Black hole explosions?" *Nature*, 248 (5443): 30–31, 1974-03-01.

Appendix D: Nature of Hamiltonian

Classical Hamiltonian was developed from an empty space platform, as shown in Figure D.1 as given by:

$$\mathcal{H} = -(1/2)mv^2 + V \geq 0$$

where m is the mass and v is the velocity. Equivalently, Hamiltonian can be written in term of momentum as given by:

$$\mathcal{H} = -p^2/(2m) + V$$

where p is the particle's momentum. Strictly speaking, Hamiltonian should present as vector forms, since kinetic and potential energy are directional. As an aside. it is not a physically realizable subspace, it is an energy zero-summed energy subspace since it allows anti-matter to exist.

Nevertheless, temporal (t > 0) can be developed within a physically realizable paradigm, as depicted in Figure D.2. From which a temporal (t > 0) Hamiltonian operator limited by quantum dynamics takes the form:

$$\mathcal{H} = [h^2/(8\pi^2 m)]\nabla^2 + V, \quad t > 0$$

where t > 0 denotes the equation is subjected to by temporal imposition. ∇^2 is a Laplacian operator and V is the potential energy. From which a temporal (t > 0) Hamiltonian equation is given by:

$$\mathcal{H}\psi = \{[h^2/(8\pi^2 m)]\nabla^2 + V\}\psi = E\psi, \quad t > 0$$

where E is an eigenvalue for Hamiltonian corresponds to the eigenstate of ψ, as given by:

$$\nabla^2\psi + (8\pi^2 m/h^2)(V - E)\psi = 0, \quad t > 0$$

Nevertheless, by letting E = hv, a temporal (t > 0) Schrödinger equation can be written:

$$\nabla^2\psi + (8\pi^2 m/h^2)(V - hv)\psi = 0, \quad t > 0$$

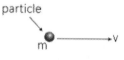

Classical Paradigm
(a piece of paper)
Empty space
It is a timeless (t = 0),
zero-summed energy subspace.
Not physically realizable paradigm

FIGURE D.1

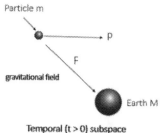

Temporal (t > 0) subspace
A physically realizable paradigm

FIGURE D.2

Since quantum leap radiated energy E = hv has no mass, the temporal (t > 0) Schrodinger equation can be reduced to:

$$\nabla^2\psi - (8\pi^2 mv/h) = 0, \quad t > 0$$

Or

$$(\partial^2\psi/\partial x^2) - (8\pi^2 mv/h) = 0, \quad t > 0$$

By reference, every physical quantum radiator is time and band limited, from which we see that E = hv, which is equivalent to the total quantum leap energy as given by:

$$\Delta t \ \Delta E = h$$

Or equivalent to:

$$\Delta t \ \Delta v = 1$$

where Δv is the quantum leap bandwidth. This is a time-bandwidth conservation of each quantum state. From this we see that a physically realizable quantum wave function can be written by:

$$\psi(t) \approx \exp[-\alpha(t - t')^2]\cos(2\pi vt), \quad t > 0$$

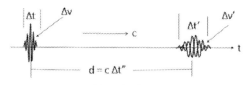

Temporal (t > 0) subspace

FIGURE D.3

$$\psi(t) = 0, \quad t < 0$$

which is a Gaussian-distributed quantum wavelet centered at t- t', where α and t' are arbitrary constants.

Since every physical aspect within our university must be temporal (t > 0), every amount of energy ΔE changes with time is written by:

$$\Delta t \ \Delta E(t) = h$$

which is an energy conservation equation since total energy is conserved with time. In other words, $\Delta E(t)$ degrades naturally with time, but overall $\Delta t \ \Delta E = h$ does not.

For example, a time-limited Δt quantum wavelet, similar to a pulse radar, is travelling at the speed of light, as depicted in Figure D.3. Since bandwidth Δv decreases as a wavelet propagates (i.e., $\Delta v' < \Delta v$ at d = c $\Delta t''$), pulse-width increases, as given by $\Delta t' \ \Delta v' = 1$. From this we see that the quantum leap energy is preserved (i.e., $\Delta t' \ \Delta E' = h$) as the wavelet propagates.

Appendix E: Why Mathematical Physicists are so Wrong?

I had found that mathematical physicists had inadvertently used a four-dimensional spacetime continuum as depicted in Figure E-1, since Einstein disclosed his Minkowsky-Lorentz spacetime over a century ago in the 1905 [1]. Since all the theoretical physicists had had entrenched with this 4-d spacetime paradigm for their theoretical analyses, which is precisely the reason scores of principles and theories were so weird and so wrong. The reason that theoretical physicists were committed this vital error was basically due to over depending on mathematics since they believe that mathematics can solve all the physical problems. On contrary, mathematics is not equaled to science although theoretical physicists need mathematics. As I see it, if any new postulated science is not developed within a physically realizable paradigm, very likely their solution will be physically unrealizable.

Since our universe is a time dependent stochastic energy conserved subspace, we see that the commonly used 4-d spacetime continuum is a zero-summed universe, in terms of energy as well in time. Beside it is not a physically realizable as from temporal (t > 0) exclusive principle (i.e., substance and emptiness are mutually

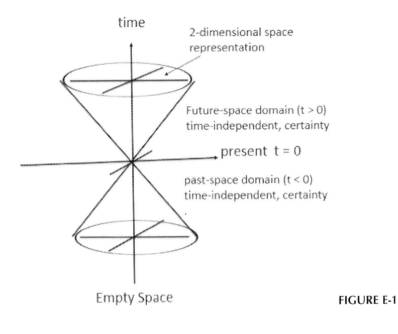

FIGURE E-1

exclusive). For which it violates the second law thermodynamics since entropy within the 4-d spacetime does not increase naturally with time. Since zero-summed universe allows antimatter to exist which is contrasting with our universe as an energy conserved universe that antimatter cannot be existed.

Figure E-1 shows a commonly used 4-d spacetime continuum. This spacetime continuum is essentially developed from the Minkowski-Lorentz transform algorithm. Where we see that time had been treated as an independent variable, in which we see that it is a zero-summed subspace. Nevertheless, it is a subspace that should "not" be used since our universe is an energy conserved subspace.

From which we see that this 4-d spacetime continuum should not had had been used for any postulated science, which includes Dirac's antimatter principle [2]. For example if Dirac had had used his equation in a non-zero summed form as given by,

$$i\partial\!\!\!/\psi = m\psi^{(i\partial\!\!\!/ - m)\psi=0}$$

he might not have had postulated his antimatter hypothesis, where ∂ is a Feynman slash notation, m is the mass of a particle and ψ is the Schrödinger wave equation. From which we see that, Dirac's equation had had missed lead us to believing that anti-particle travelling backward in time can actually exist within our temporal $(t > 0)$ universe. Although experimentally had had proven by cloud chamber experiment as reported by Anderson [3], but my question is that how we can see substance travelling backward in time within a temporal $(t > 0)$ cloud chamber. With reference to our sky is a typical cloud chamber, but we had never seen an anti-aircraft traces in negative time. If we do, this tells us that we can physically see the yesterday of ourselves. The notion of accepting antiparticle within our universe is virtual and fictitious as mathematics is. To me it seemed like searching a timeless $(t = 0)$ or backward-time particle within our temporal $(t > 0)$ universe.

Nevertheless, our temporal $(t > 0)$ universe, it is an energy conservation subspace, instead of a zero-summed energy subspace as presented by the 4-d spacetime continuum shown in the Figure E-2. For which we see that entropy within our temporal $(t > 0)$ universe increases naturally with time. And this exactly why our universe changes naturally with time, but our universe cannot change time. From all these apparent reasons as I had described, I have finally found a more convincing reason to show why current mathematical physicists are so wrong. This is precisely why relativity theories, superposition principle, anti-matter hypothesis, quantum electrodynamic equations, as well black hole postulation, and many others are so weird; fictitious as mathematics is since theoretical physicists are also mathematicians.

Figure E-2 Shows how dynamics that our universe was created. Since time is coexisted with space, we see that our universe is an energy conserved expanding subspace. And it is not a zero-summed subspace as the commonly accepted 4-d spacetime universe. From which we see that science cannot be absolutely deterministic.

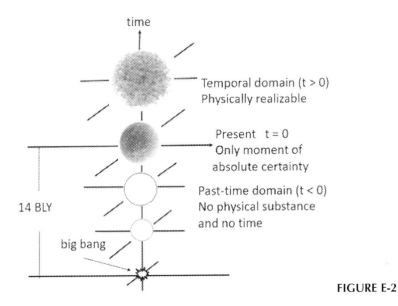

time

Temporal domain (t > 0)
Physically realizable

Present   t = 0
Only moment of
absolute certainty

Past-time domain (t < 0)
No physical substance
and no time

14 BLY

big bang

FIGURE E-2

In view of all those impotencies of mathematical physics, it must be our responsibility to let the science community knows where we were and where we are, otherwise, we will be forever trapped within a non-physically realizable subspace. By the way, it is terribly wrong to continuingly promoting all those fancy sciences which are not actually existed, since they are more harmful than benefit to science. For examples such as the nonexistent quantum qubit information, wormhole time travelling, antimatter annihilation energy, timeless angle particles, and others. In view of our universe is not a zero-summed subspace, anything within our universe has a price to pay; a section of time Δt and an amount of energy ΔE, where Δt and ΔE are coexisted.

REFERENCES

1. A. Einstein, *Relativity, the Special and General Theory*, Crown Publishers, New York, 1961.
2. P. A. M. Dirac, "On the Theory of Quantum Mechanics". *Proceedings of the Royal Society A.*, 112 (762): 661–677, (1926).
3. C. D. Anderson "The Positive Electron", *Physical Review*, 43 (6): 491–494, (1933).

Appendix F: Interpretation of Energy Equation and Wave Function

Since mass is equivalent to energy as given by Einstein's energy equation:

$$E = mc^2$$

for which we had accepted a nonzero-sum equation, where time is assumed a forward variable ($t > 0$).

However, if the same equation is presented in zero-sum forms as given by, respectively:

$$E - mc^2 = 0$$

$$mc^2 - E = 0$$

these equations are basically the same mathematically as the energy equation. But from a physical standpoint, they can be interpreted that either negative energy or negative mass (e.g., antimatter) has existed within our universe.

However, our universe is an energy or mass conservation subspace, which cannot have anti-matter or negative energy within it, as in contrast with Einstein's space-time continuum.

In view of Dirac's equation as given by:

$$(i\partial\!\!\!/ - m)\psi = 0$$

It is a zero-sum mass equation, which is precisely the reason why Dirac hypothesized anti-particle exists within our universe. If Dirac's equation is written as mass (i.e., energy) conserved subspace as given by:

$$i\partial\!\!\!/\psi = m\psi$$

He might not have had hypothesized an anti-particle, since Dirac's equation, as most of the modern physics equations, was derived from Einstein's space-time continuum. But the 4-D space-time continuum is a virtual zero-sum subspace, which should not had have been used for any physical analysis. And this is the critical issue that all modern physics is so weird and is so wrong, includes Feynman's quantum electrodynamics, string theory, Hawking's black hole,

Schrodinger's fundamental principle, Einstein's general theory, and so on of many others. And this is precisely why modern physics is so weird. From this we see that it is not how severe mathematics is, but it is the physically realizable paradigm that determines the physically realizable solution.

Since every law and equation must be temporal $(t > 0)$ (i.e., co-exist with time) to exist within our universe, from this we found Schrödinger's wave function is not a physically realizable equation. And this is precisely why modern physics needs to be revised; otherwise, we will forever be trapped within the 4-D space-time continuum of a mathematical fantasy land of science.

From this we see that particle-wave dynamics does not mean a particle in motion is equal to wave dynamics. The double slits profile distribution does not represent the probability of particle (i.e., photon) distribution, since every particle probabilistic distribution within our temporal $(t > 0)$ universe changes with time [i.e., $p(t)$]. The irradiant profile as obtained from the double-slit hypothesis is the irradiance of two coherent wavelet distributions. It does not represent the so-called photon (i.e., wave-particle dynamics) distribution but the intensity spatial distribution from the two coherent wavelets. This has nothing to do with photonic-particle distribution, since a photon is not a particle but a package of bounded quantum leap energy.

Index

191

Printed in the United States
by Baker & Taylor Publisher Services